Data Warehousing

Dedicated to

My wife, Seema, and our 10-month-old son, Vignesh, who have stuck with me through the best part of this book when I have been totally involved in the book; they received less of my attention than they ideally should have been given. Also, I dedicate this book to my parents and my in-laws, who have always been a source of inspiration to me. Their constant encouragement made me reach where I am today.

Also, I thank my friends, relatives, and colleagues who have provided input through discussions in official or personal forums, thus giving me more knowledge than what I had when I started writing this book.

Data Warehousing

By Amitesh K. Sinha

PROMPT® PUBLICATIONS

PROMPT® Publications is an imprint of Sams Technical Publishing, 5436 W. 78th St., Indianapolis, IN 46268.

International Standard Book Number: 0-7906-1249-6

Library of Congress Catalog Card Number: 2001093717

PROMPT® Publications Manager: Deborah Abshier

Acquisitions Editor: Alice J. Tripp

Senior Editor: Kim Heusel

Editor: Kelly D. Dobbs

Interior Design: Debbie Berman

Indexing: Kim Heusel

Cover Design: Christy Pierce

Graphics Conversion: Christy Pierce, Phil Velikan

Illustrations: Courtesy the author

Trademark Acknowledgments:

PRINTED IN THE UNITED STATES OF AMERICA

9 8 7 6 5 4 3 2 1

Contents at a Glance

Contents

CHAPTER 3

Issues Involved in Data Warehousing 27

CHAPTER 4

Explanation of the Star- and Snowflake Schemas 37

CHAPTER 5

A Word on Data Mining 55

CHAPTER 16

The Next Big Thing for Data Warehouses 237

CHAPTER 17

Various Tools in Data Warehousing 247

CHAPTER 18

Building a Data Warehouse—
A Few Salient Points 249

Preface

Data warehousing – a breakthrough in large-scale data storage and analysis – is ranked among the most important areas of computing today. Among all the databases in the world, Oracle is the leading RDBMS in today's technology sector. Sybase, Informix, Teradata, and Redbrick are among the leaders in data warehouse technology.

This book is a concise and precise description of a real-time data warehouse, and it dwells on the real-time experience of the author in the data warehousing industry. All the concepts that make up a data warehouse are discussed, some in greater detail than others.

Every time that you look up a book on a Web site or in a bookstore, you can find a glut of books on various technologies including data warehousing. Most books dwell on the more complex issues of data warehousing, thereby immersing you in a glut of information through which you have to sift and sort to get the details you want. This book tries to give you a firsthand view of a data warehouse, the feel of a data warehouse, and the techniques to develop a good data warehouse and data mart. Examples and portrayals from real-time projects provide readers with lots of current ideas on problems faced by project managers and developers alike while working on data warehouses.

Most books dwell too much on terminology, and readers encounter "so many words, and so many pages to read."

Acknowledgments

In addition to my family members, I would like to thank some more people who made the completion of this book possible. Undertaking a major writing project while simultaneously working long hours in an office resulted in many family sacrifices.

I have received invaluable technical advice by friends and other knowledgeable people in this area. Since this field of technology is still new, not many people understood the scope of my book but were just as ready to help me with it.

A special thanks to my company, GlobalCynex Inc., for hiring me and also giving me the encouragement to work on such a huge project. Special thanks to the president of the company, Kishore Putta, who constantly encouraged me to go on with the work.

Last, but not least, I received help from a few Web sites where I viewed a lot of material published on this subject. A couple of these Web sites were www.dmreview.com and www.datawarehousing.org, where I was able to acquire much good information on the subject. It would be a matter of collating lot of information from the Web site and putting it together for people who have an interest in this subject and who would like to know more.

Alice Tripp, acquisitions editor at Sams Technical Publishing, was first involved with this book, and I convey my thanks to her for her initial support. Also, thanks to Debbie Abshier, Prompt Publications Manager, who has been a great help in getting this book out. Thanks also to the editorial team at Sams, especially Christy Pierce and Kelly Dobbs for all their efforts, and Senior Editor Kim Heusel.

Finally, a special hello to my son, Vignesh, because of whom I could stay up late hours!!!

CHAPTER 1

Data Warehousing

What Is It? What Does It Mean? What Does It Take To Build a Data Warehouse?

Topics in this chapter:

- Introduction

- What Is Data Warehousing?

- Data Warehousing Concepts

The data warehouse is the center of this century's information systems' architecture. Data warehousing supports informational processing by providing a solid platform of integrated, historical data from which to do analysis. The data warehouse provides an integration facility in a world of unintegrated application systems. Data warehousing is achieved in an evolutionary, one-step-at-a-time fashion. The data warehouse organizes and stores the data needed for informational, analytical processing over a long period of time. A world of promise exists in building and maintaining a data warehouse.

INTRODUCTION

In today's competitive global business environment, understanding and managing enterprise-wide information is crucial for making timely decisions and responding to changing business conditions. Many companies are realizing a business advantage by leveraging one of their key assets – business data. A tremendous amount of data is generated by business's day-to-day operational applications. In addition, a valuable data is available from external sources, such as market research organizations, independent surveys, and quality testing labs. Studies indicate that the amount of data in a given organization doubles every five years. Data warehousing has emerged as an increasingly popular and powerful concept for applying information technology to turn these huge islands of data into meaningful information for better business decisions. Meta Group, Inc., a leading consultant in the data warehousing environment, suggests that more than 90 percent of the Fortune 2000 businesses will put a data warehouse into place by the end of 2005.

WHAT IS DATA WAREHOUSING?

According to Bill Inmon, known as the father of data warehousing, a data warehouse is a subject-oriented, integrated, time-variant, nonvolatile collection of data in support of management decisions.

- *Subject-oriented data* – All relevant data about a subject is gathered and stored as a single set in a useful format.

- *Integrated data* – Data is stored in a globally accepted fashion with consistent naming conventions, measurements, encoding structures, and physical attributes, even when the underlying operational systems store the data differently.

- *Nonvolatile data* – The data warehouse is read-only; data is loaded into the data warehouse and accessed there.

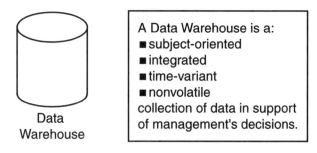

FIGURE 1.1.

What is a database warehouse?

> *Time-variant data* – This long-term data is from five to 10 years as opposed to the 30 to 60 days of operational data.

The concept of a data warehouse is depicted in Figure 1.1.

- subject-oriented

- integrated

- time-variant

- nonvolatile

This is a collection of data in support of management's decision-making process, as shown in Figure 1.1.

The data entering the data warehouse comes from the operational environment in almost every case. The data warehouse is always a physically separate store of data, which is transformed from the application data found in the operational environment.

Some of the important issues and subtleties underlying the characteristics of a data warehouse are discussed in the following sections.

Subject-Oriented

The data warehouse is oriented around the major subjects of the enterprise. The data-driven subject orientation is in contrast to the more classical process and functional orientation of applications, around which most older operational systems are organized. Figure 1.2 shows the contrast between the two types of orientations.

The operational world is designed around applications and functions, such as loans, savings, bank cards, and trusts for a financial institution. The data warehouse world is organized around major subjects, such as customer, vendor, product, and activity. The alignment around subject areas affects the design and implementation of the data found in the data warehouse. Most prominently, the major subject areas influence the most important part of the key structure.

The application world is concerned both with database design and process design. The data warehouse world focuses on data modeling and database design exclusively. Process design (in its classical form) is not part of the data warehouse environment.

The differences between process and functional application's orientation and subject orientation also show up as a difference in the content of data at the detailed level. Data warehouse data excludes data that will not be used for DSS processing, and operational application-oriented data contains data to satisfy immediate functional and processing requirements that may or may not be of use to the decision-support systems (DSS) analyst.

Another important way in which the application-oriented operational data differs from data warehouse data is in the relationships of the data. Operational data maintains an ongoing relationship between two or more tables, based on a business rule that is in effect. Data warehouse data spans a spectrum of time, and the relationships found in the data warehouse are many. Many business rules (and correspondingly, many data relationships) are represented in the data warehouse between two or more tables.

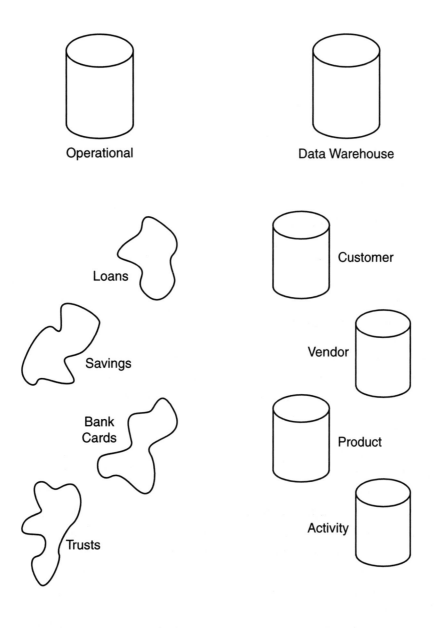

FIGURE 1.2.

The data warehouse has a strong subject orientation.

From the perspective of the fundamental difference between an application orientation and a subject orientation, a major difference exists between operational systems and data and the data warehouse.

Integrated

Easily the most important aspect of the data warehouse environment is that data found within the data warehouse is integrated – ALWAYS, WITH NO EXCEPTIONS. The very essence of the data warehouse environment is that the data contained within the boundaries of the warehouse is integrated.

The integration shows up in many different ways – in consistent naming conventions, in the consistent measurement of variables, in consistent encoding structures, in the consistent physical attributes of data, and so forth.

Contrast the integration found within the data warehouse with the lack of integration found in the applications' environment, and the differences are stark, as shown by Figure 1.3.

Over the years, the different application designers have made numerous individual decisions as to how an application should be built. The style and the individualized design decisions of the application designer show up in a hundred ways: In differences in encoding, in differences in key structures, in differences in physical characteristics, in differences in naming conventions, and so forth. The collective ability of many application designers to create inconsistent applications is legendary. Figure 1.3 shows some of the most important differences in the ways applications are designed.

> *Encoding* – Application designers have chosen to encode the field GEN-
> DER in different ways. One designer represents GENDER as an
> "M" and an "F." Another application designer represents GENDER
> as a "1" and a "0." Another application designer represents GEN-
> DER as an "x" and a "y." And, yet another application designer
> represents GENDER as "male" and "female." It doesn't matter

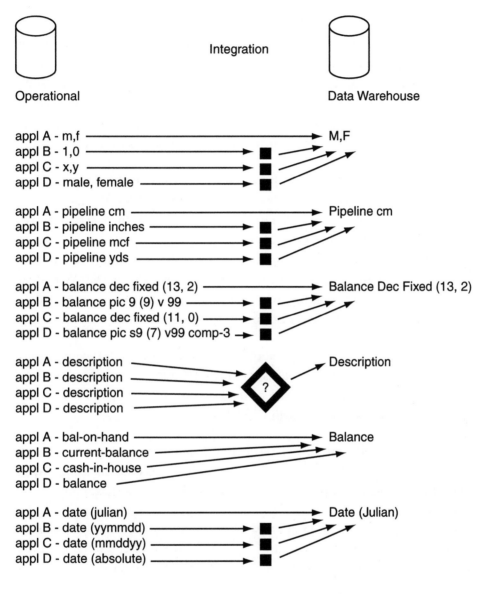

FIGURE 1.3.

When data is moved to the data warehouse from the application-oriented
operational environment, the data is integrated before
entering the warehouse.

much how GENDER arrives in the data warehouse; "M" and "F" are probably as good as any representation. What matters is that from whatever source GENDER comes, it must arrive in the data warehouse in a consistent integrated state. Therefore, when GENDER is loaded into the data warehouse from an application where it has been represented in other than an "M" and "F" format, the data must be converted to the data warehouse's format.

Measurement of attributes – Application designers have chosen to measure pipeline in a variety of ways over the years. One designer stores pipeline data in centimeters. Another application designer stores pipeline data in terms of inches. Another application designer stores pipeline data in million cubic feet per second. And, another designer stores pipeline information in terms of yards. Whatever the source, when the pipeline information arrives in the data warehouse, it needs to be measured the same way.

As shown in Figure 1.3, the issues of integration affect almost every aspect of design – the physical characteristics of data, the dilemma of having more than one source of data, the issue of inconsistent naming standards, inconsistent date formats, and so forth.

Whatever the design issue, the result is the same: The data needs to be stored in the data warehouse in a singular, globally acceptable fashion even when the u nderlying operational systems store the data differently.

When the DSS analyst looks at the data warehouse, the focus of the analyst should be on using the data that is in the warehouse, rather than on wondering about the credibility or consistency of the data.

Time Variancy

All data in the data warehouse is accurate as of some moment in time. This basic characteristic of data in the warehouse is very different from data found in the operational environment. In the operational environment, data is accurate as of the moment of access. In the operational environment, when you access a unit of data, you expect that it will reflect accurate values as of the moment of access.

Because data in the data warehouse is accurate as of some moment in time (for example, not right now), data found in the warehouse is said to be *time variant.* Figure 1.4 shows the time variancy of data warehouse data.

The time variancy of data warehouse data shows up in several ways. The simplest way is that data warehouse data represents data over a long time period — from five to 10 years. The time horizon represented for the operational environment is much shorter — from the current values of today up to 60 to 90 days. Applications that must perform well and must be available for transaction processing must carry the minimum amount of data if they are to have any degree of flexibility. Therefore, operational applications have a short time horizon, as a matter of sound application design.

Operational

Data Warehouse

Current Value Data:
- Time horizon 60-90 days.
- Key may or may not have an element of time.
- Data can be updated.

Snapshot Data:
- Time horizon 5-10 years.
- Key contains an element of time.
- Once snapshot is made, record cannot be updated.

FIGURE 1.4.

Time variancy.

The second way that time variancy shows up in the data warehouse is in the *key structure.* Every key structure in the data warehouse contains, implicitly or explicitly, an element of time, such as day, week, month, and so on. The element of time is almost always at the bottom of the concatenated key found in the data warehouse. On occasions, the element of time will exist implicitly, such as the case in which an entire file is duplicated at the end of the month or the quarter.

The third way that time variancy appears is that data warehouse data, after correctly recorded, cannot be updated. Data warehouse data is, for all practical purposes, a long series of snapshots. Of course, if the snapshots of data have been taken incorrectly, the snapshots can be changed. Assuming that snapshots are made properly, however, they are not altered once made. In some cases it may be unethical or even illegal for the snapshots in the data warehouse to be altered. Operational data, being accurate as of the moment of access, can be updated as the need arises.

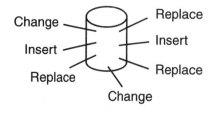

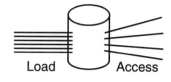

Operational

Data is updated on a record-by-record basis regularly.

Data Warehouse

Data is loaded into the warehouse and is accessed there, but after the snapshot of data is made, the data in the warehouse does not change.

FIGURE 1.5.

Nonvolatile data.

Nonvolatile

The fourth defining characteristic of the data warehouse is that it is nonvolatile. Figure 1.5 illustrates this aspect of the data warehouse.

Figure 1.5 shows that updates — inserts, deletes, and changes — are done regularly to the operational environment on a record-by-record basis. The basic manipulation of data that occurs in the data warehouse, however, is much simpler. Only two kinds of operations occur in the data warehouse — the initial loading of data and the access of data. No update of data (in the general sense of update) occurs in the data warehouse as a normal part of processing.

This basic difference between operational processing and data warehouse processing results in some very powerful consequences. At the design level, the need to be cautious of the update anomaly is no factor in the data warehouse, because data updating is not done. Therefore at the physical level of design, liberties can be taken to optimize data access, particularly in dealing with the issues of normalization and physical denormalization. Another consequence of the simplicity of data warehouse operation is in the underlying technology used to run the data warehouse environment. Having to support record-by-record updating in an online mode (as is often the case with operational processing) requires the technology to have a very complex foundation underneath a facade of simplicity. The technology supporting backup and recovery, transaction and data integrity, and the detection and remedy of deadlock is quite complex and unnecessary for data warehouse processing.

The characteristics of a data warehouse — subject orientation of design, integration of data within the data warehouse, time variancy, and simplicity of data management — all lead to an environment that is *VERY* different from the classical operational environment.

The source of nearly all data warehouse data is the operational environment. One might be tempted to think that a massive redundancy of data exists between the two environments. Indeed, the first impression many people draw is that of great redundancy of data between the operational environment and the data warehouse environment. Such an understanding is superficial and demonstrates a lack of understanding as to what is occurring in the data warehouse. In fact there is a MINIMUM of redundancy of data between the operational environment and the data warehouse environment.

Consider the following:

- Data is filtered as it passes from the operational environment to the data warehouse environment. Much data never passes out of the operational environment. Only the data that is needed for DSS processing finds its way into the data warehouse environment.

- The time horizon of data is very different from one environment to the next. Data in the operational environment is very fresh. Data in the warehouse is much older. From the perspective of time horizons alone, little overlap exists between the operational and the data warehouse environments.

- The data warehouse contains summary data that is never found in the operational environment.

- Data undergoes a fundamental transformation as it passes into the data warehouse. Figure 1.3 illustrates that most data is significantly altered upon being selected for and moving into the data warehouse; most data is physically and radically altered as it moves into the warehouse. This data is not the same data that resides in the operational environment.

In light of these factors, data redundancy between the two environments is a rare occurrence, resulting in less than 1 percent redundancy between the two environments.

Data warehousing is a concept. It is a set of hardware and software components that can be used to better analyze the massive amounts of data that companies are accumulating to make better business decisions. Data warehousing is not just data in the data warehouse, but also the architecture and tools to collect, query, analyze, and present information.

DATA WAREHOUSING CONCEPTS

Operational and Informational Data

Operational data is the data you use to run your business. This data is what is typically stored, retrieved, and updated by your Online Transactional Processing (OLTP) system. An OLTP system may be, for example, a reservations system, an accounting application, or an order-entry application.

Informational data is created from the wealth of operational data that exists in your business and some external data useful to analyze your business. Informational data is what makes up a data warehouse. Informational data is typically

- Summarized operational data

- Denormalized and replicated data

- Infrequently updated from the operational systems

- Optimized for decision support applications

- Possibly read only (no updates allowed)

- Stored on separate systems to lessen the impact on operational systems

OLAP and Multidimensional Analysis

Relational databases store data in a two-dimensional format: tables of data represented by rows and columns. Multidimensional analysis solutions, or Online Analytical Processing (OLAP) solutions, offer an extension to the relational model to provide a multidimensional view of the data. For example, in multidimensional analysis, data entities such as products, geographies, time periods, store locations, promotions, and sales channels may all represent different dimensions. Multidimensional solutions provide the ability to

- Analyze potentially large amounts of data with very fast response times.

- "Slice and dice" through the data and drill down or roll up through various dimensions as defined by the data structure.

- Quickly identify trends or problem areas that would otherwise be missed.

Multidimensional data structures can be implemented with multidimensional databases or extended Relational Database Management Systems (RDBMSs). Relational databases can support this structure through specific database designs (schema), such as star-schema, which is intended for multidimensional analysis and highly indexed or summarized designs. These structures are sometimes referred to as relational OLAP (ROLAP)-based structures.

Data Marts

Data marts are workgroup or departmental warehouses, which are small in size, typically 10–50G. The data mart contains informational data that is departmentalized, tailored to the needs of the specific departmental work group. Data marts are less expensive and take less time for implementation with quick ROI (Return on Investment). They are scalable to full data warehouses and at times are summarized subsets of more detailed, preexisting data warehouses.

Metadata and Information Catalogue

Metadata describes the data contained in the data warehouse (for example, data elements and business-oriented description) as well as the source of that data and the transformations or derivations that may have been performed to create the data element.

Data Mining

Data mining predicts future trends and behaviors, allowing business managers to make proactive, knowledge-driven decisions. Data mining is the process of analyzing business data in the data warehouse to find unknown patterns or rules of information that you can use to tailor business operations. For instance, data mining can find patterns in your data to answer questions, such as the following:

- What item purchased in a given transaction triggers the purchase of additional related items?

- How do purchasing patterns change with store locations?

- What items tend to be purchased using credit cards, cash, or check?

- How would the typical customer likely to purchase these items be described?

- Did the same customer purchase related items at another time?

CHAPTER 2

Methodology for
Data Warehousing

Topics in this chapter:

- Data Warehouse Implementation

- Structure of a Data Warehouse

- Old Detail Storage Medium

- Flow of Data

DATA WAREHOUSE IMPLEMENTATION

The following components should be considered for a successful implementation of a Data Warehousing solution:

- Open Data Warehousing architecture with common interfaces for product integration

- Data modeling with the ability to model star-schema and multidimensionality

Data Warehousing Model

Data warehousing architecture model.

- Extraction and transformation/propagation tools to load the data warehouse

- Data warehouse database server

- Analysis of end-user tools: OLAP/multidimensional analysis, report and query

- Tools to manage information about the warehouse (metadata)

- Tools to manage the data warehouse environment

Transforming Operational Data into Informational Data

Creating the informational data, the data warehouse, from the operational systems is a key part of the overall data warehousing solution. Building the informational database is done with the use of transformation or propagation tools. These tools not only move the data from multiple operational systems but often manipulate the data into a more appropriate format for the warehouse. This could mean the following:

- The creation of new fields that are derived from existing operational data

- Summarizing data to the most appropriate level needed for analysis

- Denormalizing the data for performance purposes

- Cleansing of the data to ensure that integrity is preserved

Even with the use of automated tools, however, the time and costs required for data conversion are often significant. Bill Inmon has estimated 80 percent of the time required to build a data warehouse is typically consumed in the conversion process.

Data Warehouse Database Servers — The Heart of the Warehouse

When ready, data is loaded into a relational database management system (RDBMS), which acts as the data warehouse. Some of the requirements of database servers for data warehousing include performance, capacity, scalability, open interfaces, multiple data structures, an optimizer to support for star-schema, and bitmapped indexing. Some of the popular data stores for data warehousing are relational databases like Oracle, DB2, and Informix or specialized data warehouse databases like RedBrick and SAS.

To provide the level of performance needed for a data warehouse, an RDBMS should provide capabilities for parallel processing: Symmetric Multiprocessor (SMP) or Massively Parallel Processor (MPP) machines, near-linear scalability, data partitioning, and system administration.

STRUCTURE OF A DATA WAREHOUSE

Data warehouses have a distinct structure. Different levels of summarization and detail demark the data warehouse. The structure of a data warehouse is shown by Figure 2.1.

Figure 2.1 shows that the different components of the data warehouse are

- Metadata
- Current detail data
- Older detail data
- Lightly summarized data
- Highly summarized data

Far and away the major concern is the current detail data, because

- Current detail data reflects the most recent happenings, which are always of great interest.
- Current detail data is voluminous because it is stored at the lowest level of granularity.
- Current detail data is almost always stored on disk storage, which is fast to access but expensive and complex to manage.

Older detail data, is data, that is stored on some form of mass storage. It is infrequently accessed and is stored at a level of detail consistent with current detailed data. While not mandatory that it be stored on an alternate storage medium, because of the anticipated large volume of data coupled with the infrequent access of the data, the storage medium for older detailed data is usually not disk storage.

Lightly summarized data is data that is distilled from the low level of detail found at the current detailed level. This level of the data warehouse is

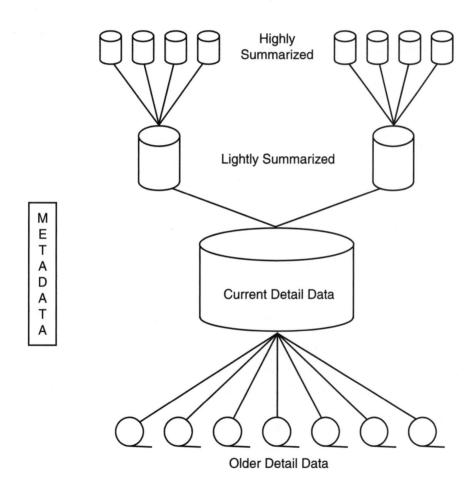

FIGURE 2.1

The structure of data inside the data warehouse.

almost always stored on disk storage. The design issues facing the data architect in building this level of the data warehouse are

- What unit of time is the summarization done over, and

- What contents – attributes – will the lightly summarized data contain

The next level of data found in the data warehouse is that of the highly summarized data. Highly summarized data is compact and easily accessible. Sometimes, the highly summarized data is found in the data warehouse environment, and in other cases, the highly summarized data is found outside the immediate walls of the technology that houses the data warehouse. (In any case, the highly summarized data is part of the data warehouse, regardless of where the data is physically housed.)

The final component of the data warehouse is that of metadata. In many ways, metadata sits in a different dimension than other data warehouse data, because metadata contains no data directly taken from the operational environment. Metadata plays a special and very important role in the data warehouse. Metadata is used as

- A directory to help the DSS analyst locate the contents of the data warehouse

- A guide to the mapping of data as the data is transformed from the operational environment to the data warehouse environment

- A guide to the algorithms used for summarization between the current detailed data and the lightly summarized data, between the lightly summarized data and the highly summarized data, and so on.

Metadata plays a much more important role in the data warehouse environment than it ever did in the classical operational environment.

In order to bring to life the different levels of data found in the data warehouse, consider the example shown in Figure 2.2.

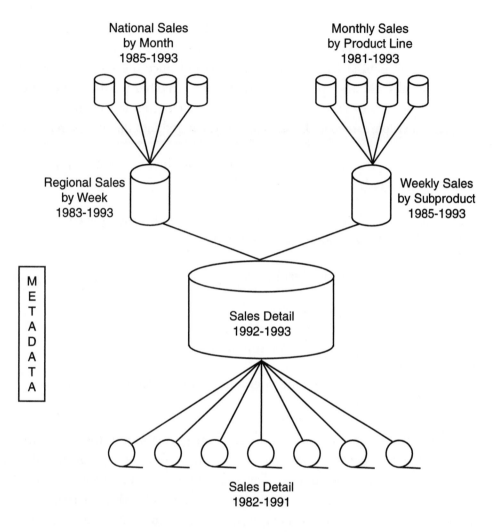

FIGURE 2.2.

An example of the levels of summarization that might be
found in the data warehouse.

In Figure 2.2 old sales detail is that detail about sales that is older than 1992.
All sales detail from 1982 (or whenever the data architect is able to start
collecting archival detail) is stored in the old detail level of data.

The current value detail contains data from 1992 to 1993 (assuming that 1993 is the current year). As a rule, sales detail does not find its way into the current level of detail until at least 24 hours have passed since the sales information became available to the operational environment. In other words, a time lag of at least 24 hours between the time the operational environment got news of the sale and the moment when the sales data was entered into the data warehouse.

The sales detail is summarized weekly by subproduct line and by region to produce the lightly summarized stores of data.

The weekly sales detail is further summarized monthly along even broader lines to produce the highly summarized data.

Metadata contains (at the least!)

- The structure of the data

- The algorithms used for summarization

- The mapping from the operational environment to the data warehouse

Note that not every summarization ever done gets stored in the data warehouse. On many occasions analysis will be done, and one type or the other of summary will be produced. The only type of summarization that is permanently stored in the data warehouse is that data which is frequently used. In other words, if a DSS analyst produces a summarized result that has a very low probability of ever being used again, then that summarization is not stored in the data warehouse.

OLD DETAIL STORAGE MEDIUM

The symbol shown in Figure 2.2 for old detail storage medium is that of magnetic tape. Indeed, magnetic tape may be used to store that type of data. A wide variety of storage media should be considered for storing older detail data. Figure 2.3 shows some of those media.

- Phot Optical Storage
- RAID
- Microfiche
- Magnetic Tape
- Mass Storage

FIGURE 2.3.

The storage medium for the bulk portion of the data warehouse can be a wide variety of storage types.

Depending on the volume of data, the frequency of access, the cost of the media, and the type of access, other storage media may serve the needs at the old level of detail in the data warehouse.

FLOW OF DATA

A normal and predictable flow of data exists within the data warehouse. Figure 2.4 shows that flow.

Data enters the data warehouse from the operational environment. (Note: A few very interesting exceptions exist to this rule. However, *nearly all* data enters the data warehouse from the operational environment.) As data enters the data warehouse from the operational environment, it is transformed, as has been described earlier.

Upon entering the data warehouse, data goes into the current detail level of detail, as shown. It resides there and is used until one of three events occurs:

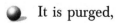 It is purged,

It is summarized, and/or

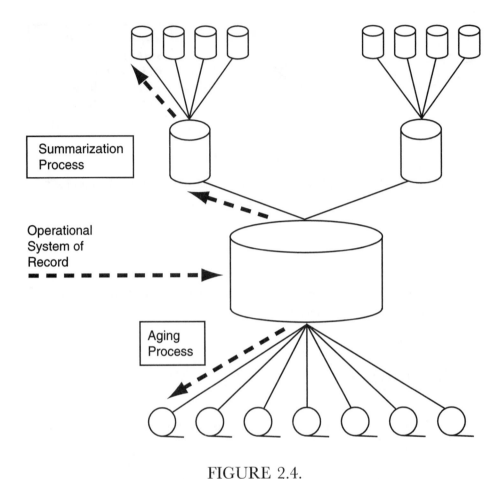

FIGURE 2.4.

The flow of data inside the data warehouse.

⬤ It is archived.

The aging process inside a data warehouse moves current detail data to older detail data, based on the age of the data. The summarization process uses the detail of data to calculate the lightly summarized data and the highly summarized data.

A few exceptions exist to the flow as shown (these exceptions will be discussed later in this book). For the vast majority of data found inside a data warehouse, however, the flow of data is as depicted in this chapter.

CHAPTER 3

Issues Involved in Data Warehousing

Topics in this chapter:

- Using the Data Warehouse

- Other Considerations

- An Example of a Data Warehouse

- Other Anomalies

- Summary

USING THE DATA WAREHOUSE

The different levels of data within the data warehouse receive different levels of usage, not surprisingly. As a rule, the higher the level of summarization, the more the data is used, as shown in Figure 3.1.

Figure 3.1 shows that much usage occurs in the highly summarized data, although the older detail data is hardly ever used.

27

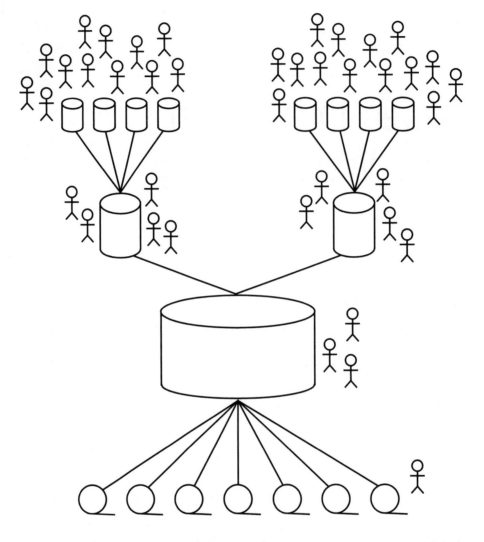

FIGURE 3.1.

The higher the levels of summarization, the more the usage of the data.

A good reason for moving the organization to the paradigm is suggested in Figure 3.1 – resource utilization. The more summarized the data, the quicker and the more efficient it is to get to the data. If a shop finds that it is doing a lot of processing at the detailed levels of the data warehouse, then a correspondingly large amount of machine resources are being consumed. It is in everyone's best interests to do processing at as high a level of summarization as possible.

For many shops, the DSS analyst in a predata warehouse environment has used data at the detailed level. In many ways, getting to detailed data is like a security blanket, even when other levels of summarization are available. One of the tasks of the data architect is to wean the DSS user from constantly using data at the lowest level of detail. Two motivators are at the disposal of the data architect:

- Installing a chargeback system, where the end user pays for resources consumed

- Pointing out that very good response time can be achieved when dealing with data at a high level of summarization, although poor response time results from dealing with data at a low level of detail.

OTHER CONSIDERATIONS

Some other issues must be considered when building and administering the data warehouse. Figure 3.2 shows some of those considerations.

The first consideration is that of indexes. Data at the higher levels of summarization can be freely indexed, although data at the lower levels of detail is so voluminous that it can be indexed sparingly. By the same token, data at the higher levels of detail can be restructured relatively easily, although the volume of data at the lower levels is so great that data cannot be easily restructured.

Accordingly, the data model and formal design work that lay the foundation for the data warehouse, apply almost exclusively to the current level of detail. In other words, the data-modeling activities do not apply to the levels of summarization, in almost every case.

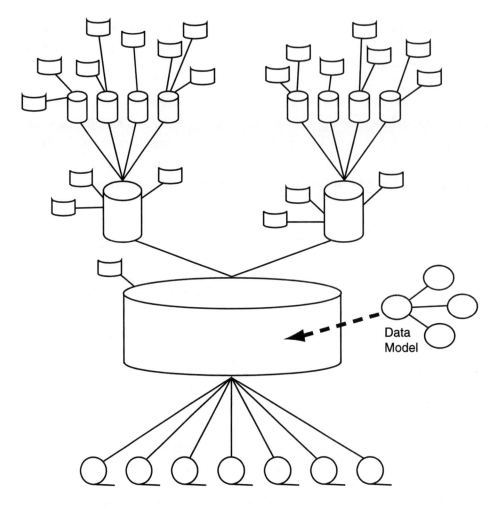

FIGURE 3.2.

The more summarized the data is, the more indexes there are, and the data model applies to the current level of detail.

Another structural consideration is that of the partitioning of data warehouse data. Figure 3.3 shows that current level detail is almost always partitioned.

Figure 3.3 shows that partitioning can be done in two ways: at the DBMS level and at the application level. In DBMS partitioning, the DBMS is aware of the partitions and manages them accordingly. In application partitioning,

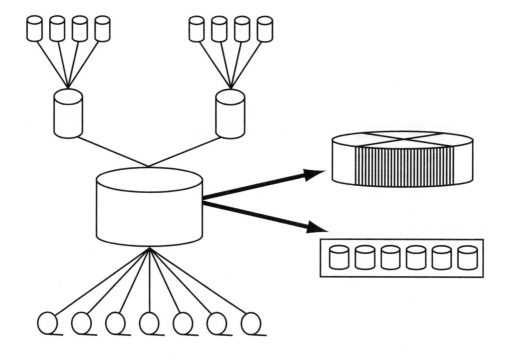

FIGURE 3.3.

Current detail data is almost always partitioned.

only the application programmer is aware of the partitions, and responsibility for the management of the partitions is left up to the programmer.

Under DBMS partitioning, much infrastructure work is done automatically. However, a tremendous degree of inflexibility is associated with the automatic management of the partitions. In application partitioning of data warehouse data, much work falls to the programmer, but the end result is greater flexibility in the management of the data.

AN EXAMPLE OF A DATA WAREHOUSE

Figure 3.4 shows a hypothetical example of a data warehouse structured for a manufacturing environment.

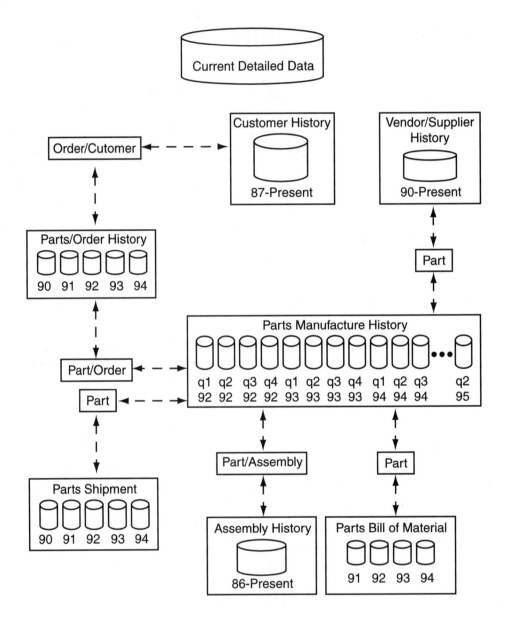

FIGURE 3.4.

The internal structuring of data in the data warehouse.

Figure 3.4 shows only current detail data. The levels of summarization are not shown, nor is the old detail archive shown.

Figure 3.4 shows that tables of the same type are divided over time. For example, for parts manufacture history, many physically separate tables are available, each representing a different quarter. The structure of the data is consistent within the parts manufacture history table, even though physically many tables logically comprise the history.

Note that different units of time physically divide the units of data in the different types of tables. Manufacturing history is divided by quarter; part/order history is divided by year; and customer history is a single file, not divided by time.

Also note that the different tables are linked by means of a common identifier – parts, parts/orders, and so on. (Note: The representation of a relationship in the data warehouse environment takes a very different form than relationships represented in other environments, such as the operational environment.

OTHER ANOMALIES

Although the data warehouse components work in the same fashion for almost all data, a few worthwhile exceptions need to be discussed. One exception is that of public summary data. Public summary data is summary data that has been calculated outside the boundaries of the data warehouse but is used throughout the corporation. Public summary data is stored and managed in the data warehouse, even though its calculation is well outside the data warehouse. A classical example of public summary data is the quarterly filings made by every public company to the Securities and Exchange Commission (SEC). The accountants work to produce such numbers as quarterly revenue, quarterly expenses, quarterly profit, and so forth. The work done by the accountants is well outside the data warehouse. However, those benchmark numbers produced by the accountants are used widely within the corporation – by marketing, sales, and so on. After the SEC filing is done, the data is stored in the data warehouse.

Another exceptional type of data sometimes found in a data warehouse is that of permanent detail data. Permanent detail data results in the need of a corporation to store data at a detailed level permanently for ethical or legal reasons. If a corporation is exposing its workers to hazardous substances, a need for permanent detail data exists; if a corporation produces a product that involves public safety, such as building airplane parts, a need for permanent detail data exists; if a corporation engages in hazardous contracts, a need for permanent detail data also exists; and so forth.

The corporation simply cannot afford to let go of details because in future years, in the case of a lawsuit, a recall, a disputed building flaw, and so on, the exposure of the company will be great. Therefore, a unique type of data in the data warehouse is known as permanent detail data.

Permanent detail data shares many of the same considerations as other data warehouse data, except that

- The medium the data is stored on must be as safety proof as possible.

- The data must be able to be restored.

- The data needs special treatment in the indexing of it; otherwise the data may not be accessible even though it has been safely stored.

SUMMARY

A data warehouse is a subject-oriented, integrated, time-variant, nonvolatile collection of data in support of management's decision needs. Each of the salient aspects of a data warehouse carries its own implications.

Four levels of data warehouse data exist:

- Old detail

- Current detail

- Lightly summarized data

- Highly summarized data

Metadata is also an important part of the data warehouse environment. Each of the levels of detail carries its own considerations.

CHAPTER 4

Explanation of the Star- and Snowflake Schemas

Topics in this chapter:

- Star-Schema

- Benefits of Data Warehousing

- Summary

STAR-SCHEMA

The simple model we will use to demonstrate the various design alternatives is composed of three dimensions (Figure 4.1). Only two are shown, Store and Product. The third, Time, is composed of the following attribute hierarchy: date $\varnothing$ month $\varnothing$ quarter $\varnothing$ year.

The Store dimension has an attribute hierarchy of store $\varnothing$ district $\varnothing$ region. Products is composed of products $\varnothing$ brand $\varnothing$ manufacturer.

OUR MODEL

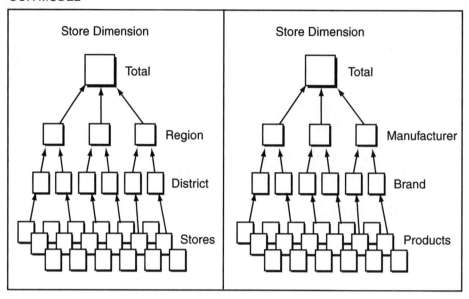

FIGURE 4.1.

Star-schema model.

Based on this simple model, we can see that the granularity of data is products sold in stores by day.

An expanded view of the model (Figure 4.2) shows three dimensions: Time, Store, and Product. Attribute hierarchies shown are vertical relationships, and extended attribute characteristics are diagonal.

In the time dimension, a given date is further described by its extended attributes "current flag," "sequence," and "day of the week." Extended attribute characteristics have no impact on granularity. The fact that February 4, 1996, is on a Sunday has no effect on the fact that we collect sales by day. In practice, though, we may want to compare Sunday sales to other days. We can do this by constraining our query on the extended attribute characteristic "day of the week" without having to gather any additional information.

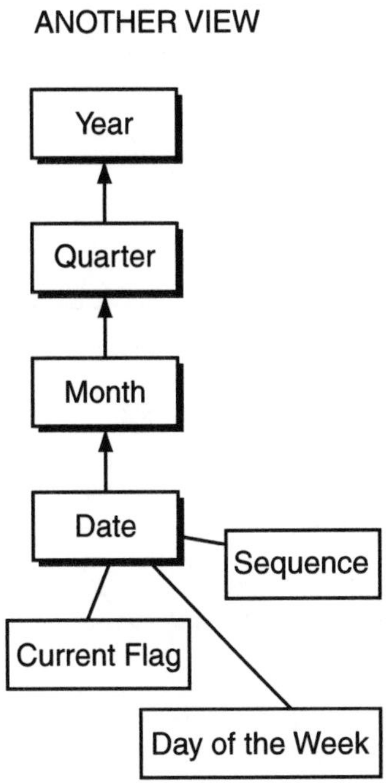

ANOTHER VIEW

FIGURE 4.2.

Another view of our model.

The Store dimension includes an extended attribute characteristic at a higher level: Each region has a Regional Manager.

Remember that the attribute hierarchies imply aggregation of data: stores roll up into districts; districts into regions. Now let's look at a relational schema for capturing these relationships (Figure 4.3).

The classic star-schema is characterized by the following:

- A single fact table containing a compound primary key, with one segment for each dimension and additional columns of additive, numeric facts

- A single dimension table (for each dimension) with a generated key, and a level indicator that describes the attribute level of each

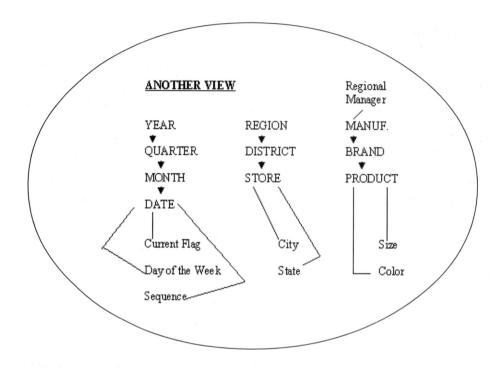

FIGURE 4.3.

The classic star-schema.

record. For example, if the dimension is Store, records in the dimension table might refer to stores, districts, or regions. We might assign a level number to these (such as 1=store, 2=district, 3=region) and put that value in the "level" column.

- The single fact table will contain detail (or *atomic*) data, such as sales dollars, for a given store, for a given product, in a given time period.

- The fact table may also contain partially consolidated data, such as sales dollars for a region, for a given product, for a given time period.

Confusion is possible if *all* consolidated data isn't included. For example, if data is consolidated only to the district level, a query for regional information

A WORD ABOUT INDEXING...
COMPOUND KEYS

STORE	DISTRICT	REGION	STORE DESCRIP	LEVEL
FL43A7	TEXA	SOUTH	PLANO#3	STORE
AR43A6	TEXA	SOUTH	ARLINGTON#2	STORE
PO12B3	EPEN	NORTH	POTTSTOWN	STORE
LA12A6	EPEN	NORTH	LANSDA.E	STORE
NULL	TEXA	SOUTH	NULL	DISTRICT
NULL	EPEN	NORTH	NULL	DISTRICT
NULL	NULL	NORTH	NULL	REGION

Compound Keys

FIGURE 4.4.

An example of compound keys.

will bring back no records and, hence, report no sales activity. For this reason, simple star-schemas *must* contain either

ALL of the combinations of aggregated data

At least views of every combination

In Figure 4.4, there is a hierarchy of attributes: store ∅ district ∅ region. One approach is to create a multipart key, identifying each record by the combination of store/district/region. Although this approach is acceptable in normalized designs, it poses some problems in the multidimensional model. For three major reasons, using compound keys in a dimension table can cause problems:

1. It requires three separate metadata definitions to define a single relationship, which adds to complexity in the design and sluggishness in performance.

2. Because the fact table must carry all three keys as part of its primary key, addition or deletion of levels in the hierarchy (such as the

addition of territory between store and district) will require physical modification of the fact table, a time-consuming process that limits flexibility

3. Carrying all of the segments of the compound dimensional key in the fact table increases the size of the crucial fact table index, a real determinant to both performance and scalability.

One alternative to compound keys is to link the keys into a single key. Although this approach solves the first two problems with compound keys (extra metadata and rigidity in the fact table), the size of the key is still a problem. Also, as in Figure 4.5, dealing with nulls can be confusing.

The best solution is to drop the use of meaningful keys and to generate the smallest possible key that will ensure the uniqueness of each record (Figure 4.6). Integers are the most efficient in most cases.

Note that the meaningful keys do not have to disappear; they may be shifted to nonkey attribute columns. If, in fact, these attributes are used

A WORD ABOUT INDEXING...
CONCATENATED KEYS

CONCATENATED STORE KEY			STORE DESCRIP	LEVEL
FL43A7	TEXA	SOUTH	PLANO#3	STORE
AR43A6	TEXA	SOUTH	ARLINGTON#2	STORE
PO12B3	EPEN	NORTH	POTTSTOWN	STORE
LA12A6	EPEN	NORTH	LANSDA.E	STORE
NULL	TEXA	SOUTH	NULL	DISTRICT
NULL	EPEN	NORTH	NULL	DISTRICT
NULL	NULL	NORTH	NULL	REGION

Concatenated Keys

FIGURE 4.5.

Concatenated keys.

frequently in queries (where region _description is "North"), the columns can still be indexed, even if they aren't used as the key.

In summary, the use of generated keys is preferred because:

- They allow for the highest level of flexibility of metadata.

- They are low maintenance as the data warehouse matures.

- They offer the highest possible performance.

The star-schema is built for simplicity and speed (Figure 4.7). Forget everything you learned about designing relational databases. The star-schema makes no excuses for the rules it breaks. The assumption behind it is that the database is static or quiet, meaning that no updates are performed online.

Remember that most of the rules of relational database design are derived from the need to maintain atomicity, consistency, and integrity (the "ACID" test) in an Online Transactional Processing (OLTP) environment. Because the data warehouse is quiet, these constraints can be relaxed.

A WORD ABOUT INDEXING...
GENERATED KEYS

STORE KEY	STORE	DISTRICT	REGION	STORE DESCRIP		LEVEL
101	FL43A7	TEXA	SOUTH	PLANO#3		STORE
102	AR43A6	TEXA	SOUTH	ARLINGTON#2		STORE
103	PO12B3	EPEN	NORTH	POTTSTOWN		STORE
104	LA12A6	EPEN	NORTH	LANSDA.E		STORE
105	NULL	TEXA	SOUTH	NULL		DISTRICT
106	NULL	EPEN	NORTH	NULL		DISTRICT
107	NULL	NULL	NORTH	NULL		REGION

Generated Keys

FIGURE 4.6.

Generated keys.

THE "CLASSIC" STAR-SCHEMA

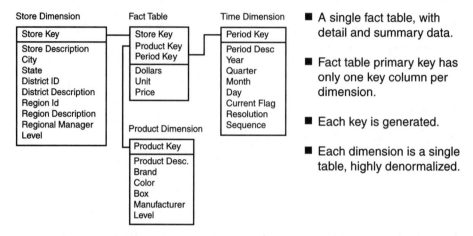

- A single fact table, with detail and summary data.

- Fact table primary key has only one key column per dimension.

- Each key is generated.

- Each dimension is a single table, highly denormalized.

Benefits: Easy to understand, easy to define hierarchies, reduces # of physical joins, low maintenance, very simple metadata.

Drawbacks: Summary data in the fact table yields poorer performance for summary levels, huge dimension tables a problem.

FIGURE 4.7.

The classic star-schema.

The biggest drawback of the level indicator is that it limits flexibility (Figure 4.8). Remember the notion of bounded dimensionality: We may not know all of the levels in the attribute hierarchies at first. By limiting ourselves to only certain levels, we force a physical structure that may change, resulting in higher maintenance costs and more downtime.

The level concept is a useful tool for very controlled data warehouses – that is, those that either have no *ad hoc* users or only those *ad hoc* users who are savvy about the database. In particular, when the results of queries are preformatted reports or extracts to smaller systems, such as data marts, the drawbacks of the level indicator are not so evident.

The chart in Figure 4.10 is composed of all of the tables from the Classic Star, plus aggregated fact tables. For example, the Store dimension is formed

The Classic Star Schema

The biggest drawback : Dimension tables must carry a level indicator for every record and every query must use it. In the example given here, without the level constraint, keys for all stores in the North Region, including aggregates for region and district will be pulled from the fact table, resulting in an error.

For example,

Select a.store_key, a.period_key, a.dollars
From Fact table a
Where a.store_key in (select store_key
                     From store_dimension
                     Where region = 'North' and level = 3);

Note : the bottom line is "level is needed whenever aggregates are stored with detail facts".

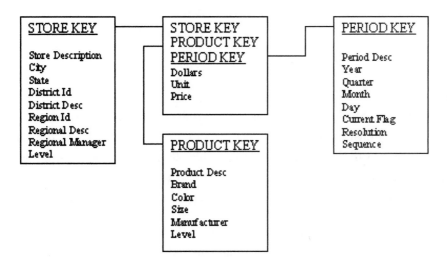

FIGURE 4.8.

The biggest drawback of the star-schema.

of a hierarchy of store ∅ district ∅ region. The base fact table contains detail data by store. The District fact table contains *only* data aggregated by district, therefore no records in the table have STORE_KEY matching any

The Level Problem

Level is a problem because it causes potential for error. If the query builder or a human or a machine program, forgets about it, a very reasonable WRONG answer can occur.

One Alternative : the FACT CONSTELLATION model.

FIGURE 4.9.

The level problem.

record for the Store dimension at the store level. Therefore, when we scan the Store dimension table and select keys that have district = "Texas," they will match STORE_KEY in the District fact table only when the record is aggregated for stores in the Texas district. No double (or triple, and so on) counting is possible, and the Level indicator is not needed (Figure 4.9).

These aggregated fact tables can get very complicated. For example, we need a District and Region fact table, but what level of detail will they contain about the product dimension? All of the following:

STORE/PROD DISTRICT/PROD REGION/PROD

STORE/BRAND DISTRICT/BRAND REGION/BRAND

STORE/MANUF DISTRICT/MANUF REGION/MANUF

And these are just the combinations from two dimensions!

The Fact Constellation Schema

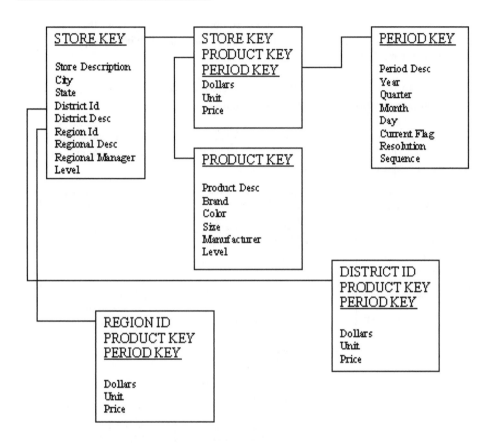

FIGURE 4.10.

Fact constellation schema.

Another drawback is that multiple SQL statements may be needed to answer a single question; for example: Measure the percent to total of a district to its region. Separate queries are needed to both the District and Region fact tables, and then some complex stitching together of the results is needed.

Once again, it is easy to see that even with its disadvantages, the Classic Star enjoys the benefit of simplicity over its alternatives (Figure 4.11).

The Fact Constellation Schema

In the Fact Constellations, aggregate tables are created separately from the detail; therefore, it is impossible to pick up, for example, store detail when querying the District Fact Table.

Major Advantage : No need for the level indicator in the dimension tables, since no aggregated data is stored with lower level detail.

Disadvantage : Dimension tables are large in some cases, which can slow the performance. The front end must be able to detect the existence of the aggregate facts, which requires more extensive metadata.

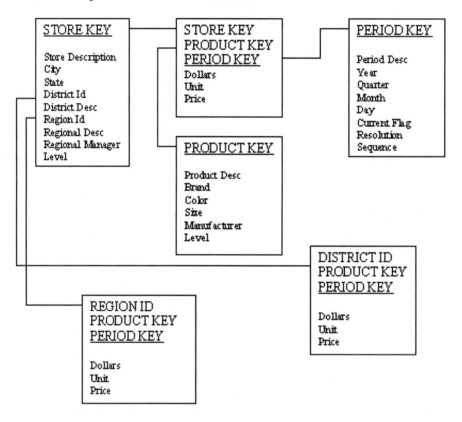

FIGURE 4.11.

Advantages and disadvantages of the fact constellation schema.

Another Alternative to Level

Fact Constellation is a good alternative to the Star Schema, but when dimensions have very high level of cardinality, the sub selects in the dimension tables can be a potential source of delay.

An Alternative is to normalize the dimension table by attribute level, with each smaller dimension table pointing to an appropriate aggregated fact table, the Snowflake Schema.

FIGURE 4.12.

An alternative to level.

Notice how the Store dimension table generates subsets of records. First, all records from the table (where level = "District" in the star-schema) are extracted, and only those attributes that refer to that level (District Description, for example) and the keys of the parent hierarchy (Region_ID) are included in the table (Figure 4.12). Although the tables are subsets, it is absolutely critical that column names are the same throughout the schema.

Figure 4.13 is a partial schema. It shows only the snowflaking of one dimension. In fact, the product and time dimensions would be similarly decomposed as follows:

Product – product ∅ brand ∅ manufacturer. (Color and size are extended attribute characteristics of the attribute "product," not part of the attribute hierarchy.)

Time – day ∅ month ∅ quarter ∅ year

The Snowflake Schema

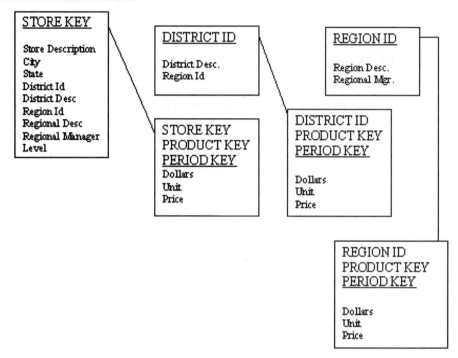

FIGURE 4.13.

The snowflake schema.

BENEFITS OF DATA WAREHOUSING

A well-designed and implemented data warehouse can be used to

- Understand business trends and make better forecasting decisions

- Bring better products to market in a more timely manner

- Analyze daily sales information and make quick decisions that can significantly affect your company's performance

Data warehousing can be a key differentiator in many different industries. At present, some of the most popular data warehouse applications include

The Snowflake Schema

⇑ There is no level in the dimension table
⇑ Dimension tables are normalized by decomposing at the attribute level.
⇑ Each dimension table has only one key for each level of the dimension's hierarchy.
⇑ The lowest level key joins the dimension table to both the fact table and the lower level attribute table.

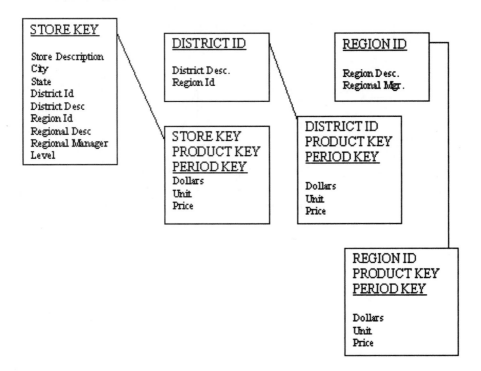

FIGURE 4.14.

The operation of the snowflake schema.

● Sales and marketing analysis across all industries

● Inventory turn and product tracking in manufacturing

● Category management, vendor analysis, and marketing program effectiveness analysis in retail

The Snowflake Schema

Features :
- The original Store Dimension table, completely denormalized, is kept intact, since certain queries can benefit by its content.
- In real life, start with a Star Schema, and create the snowflakes, with queries. This eliminates the need to create separate extracts for each table, and referential integrity is inherited from the dimension table.
- The queries best perform when they involve aggregation
- Complicated maintenance and metadata, explosion in the number of tables in the database.

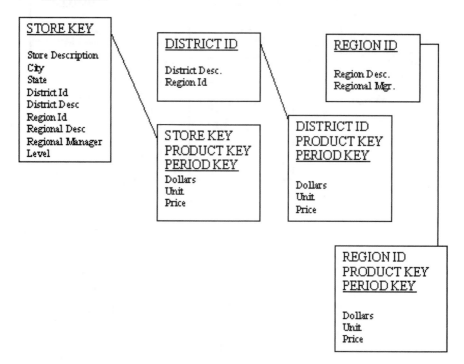

FIGURE 4.15.

Advantages and disadvantages of the snowflake schema.

- Profitable lane or driver risk analysis in transportation

- Profitability analysis or risk assessment in banking

- Claims analysis or fraud detection in insurance

SUMMARY

Data warehousing provides the means to change raw data into information for making effective business decisions – the emphasis on information, not data. The data warehouse is the hub for decision support data. A good data warehouse will provide ... the *right* data ... to the *right* people ... at the *right* time: RIGHT NOW! While the data warehouse organizes data for business analysis, the Internet has emerged as the standard for information sharing. So the future of data warehousing lies in accessibility from the Internet. Successful implementation of a data warehouse requires a high-performance, scaleable combination of hardware and software that can integrate easily with existing systems, so customers can use data warehouses to improve their decision-making and their competitive advantage.

CHAPTER 5

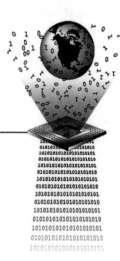

A Word on
Data Mining

Topics in this chapter:

INTRODUCTION

In the early days of data warehousing, data mining was viewed as a subset of the activities associated with the warehouse. But today, the paths of

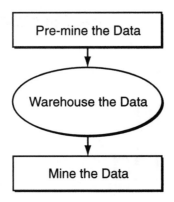

FIGURE 5.1.

Sandwiching the data.

warehousing and mining are diverging. Although a warehouse may be a good source for the data to be mined, data mining has been recognized as a bona fide task in its own right and is no longer regarded as a colony of the warehouse.

In fact, not only has data mining gained independence, but it is directly and significantly influencing the design and implementation of large data warehouses. In the early days, the "build a warehouse first, mine later" paradigm seemed simple and intuitive and, in many cases, was followed by default. A much better way is to *sandwich* the warehousing effort between two layers of mining – thus understanding the data before warehousing it, as shown in Figure 5.1.

Although warehousing and mining are undoubtedly related activities and can reinforce each other, data mining requires different data structures and computational processes and caters to a different group of users than the typical warehouse. We need to carefully separate these processes and understand how they differ in order to use them effectively.

In this chapter, I follow the path started in the Sandwich Paradigm farther and discuss how the data to be mined is placed within The Data Mine, a

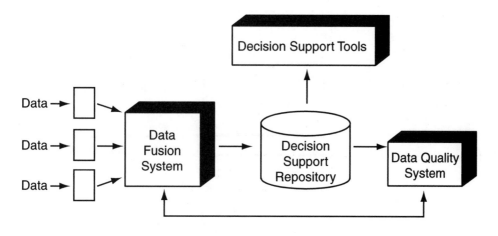

FIGURE 5.2.

DSS structure.

repository that can either coexist with or be distinct from the data warehouse. The Sandwich Paradigm then acts as a design approach that ensures that the warehouse and the mine work in unison.

STORING DATA AND ANALYZING INFORMATION

The purpose of most warehouses is to bring together large amounts of historical data from several sources and to use that data for decision support. The overall top-level structure of a decision-support system is shown in Figure 5.2.

Note that I have used the term repository rather than warehouse in Figure 5.2, because later we will see that the repository may have distinct components, such as the warehouse and the data mine. The activities performed on a large corporate data repository are usually diverse but often include distinct tasks, such as query and reporting, multidimensional analysis, and data mining. These tasks naturally break into separate user groups, as well as distinct computational processes.

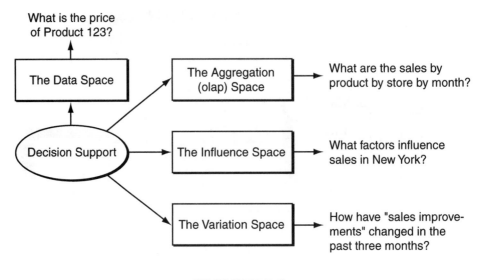

FIGURE 5.3.

The four spaces that form the basis of decision support.

Access and analysis work on different computational spaces – the more informational spaces being *derived spaces.* Data access operations, such as query and reporting, deal with the data space. OLAP uses the multidimensional space, and data mining takes place on the influence space. The four spaces that form the basis of decision support are shown in Figure 5.3. They are the spaces for data, aggregation, influence, and variation. A fifth space based on geographic relationships also may be used for some analyses.

A data warehouse is thus the natural place for storing the *data space.* It is where we store base-level data elements that are later analyzed to deliver information. And, just as OLAP is no longer viewed as a pure warehousing effort, a data mine is where we perform analyses to deal with the influence space.

The questions posed to these spaces are inherently different. Some questions, such as "What influences sales?", are almost impossible to answer directly from the data space. Moreover, the derived spaces are often so large that they can not be fully precomputed and stored like the data space;

for example, we cannot easily precompute all influence factors within a large database beforehand. Hence, we often have to rely on a partial precomputation of these spaces, with further dynamic computations performed for providing actual answers.

The data space includes all the information contained in the other spaces but in less refined form. It forms the basis for the derivation of the other spaces. However, after computed, the data space becomes individual real spaces in their own right. Hence, we start with the data space and derive the other spaces from it. We can go further and derive hybrid spaces (for example, influence analysis on aggregations and/or variations). These hybrid spaces deliver highly refined and usable information from the raw data stored within the warehouse.

In the process of analysis, the data space often needs to be enriched with additional semantics by adding information about hierarchies and periodic behavior. These additional semantics go a little beyond the simple relational model but they often help the user. For instance, {month, quarter, year} and {city, state, country} form natural hierarchies that the simple relational model ignores. By adding such semantics to the OLAP space, we provide a good deal of additional benefit to the user.

Although the OLAP space mostly deals with the computation of numeric values, the influence space has a logical nature. This space deals with the influence of specific groups of items on the others. What makes this space most interesting is that the information it provides is potentially much more useful than the other spaces, because such information is typically very general and may be referred to as *knowledge*. Because information is now one of the most valuable commodities in the world, and with the increasing complexity of today's society, the information obtained by data mining can be exponentially more valuable than any other asset.

Note that the size of the influence space and the number of logical combinations of influence factors, however, can be extremely large, making it very hard to precompute this space. Restricting this to a smaller space does not help because the goal is to find unexpected patterns, and the four fields that are excluded may just hold the key to the problem! Therefore, discovery is inherently a very dynamic process. Moreover, discovery often needs to be performed dynamically because once some unexpected item has been discovered, the user will soon begin to think of other items of interest for the program to pursue.

EXPLORATORY AND CONFIRMATORY ANALYSES

Data mining is defined as a *decision support process in which we search for patterns of information in data.* This search may be done by the user, for example just by performing queries (in which case it is quite hard) or may be assisted by a smart program that automatically searches the database by itself and finds significant patterns. This process is called *discovery.*

Discovery is the process of looking in a database to find hidden patterns without a predetermined idea or hypothesis about what the patterns may be. In other words, the program takes the initiative in finding what the interesting patterns are, without the user thinking of the relevant questions first. In large databases, so many patterns exist that the user can never think of the right questions to ask. This aspect distinguishes the space from the others.

In fact, from a statistical point of view, two types of analyses exist: *Confirmatory Analysis* and *Exploratory Analysis.* In confirmatory analysis, one has a hypothesis and either confirms or refutes it through a line of inferential statistical reasoning. However, the bottleneck for confirmatory analysis is the shortage of hypotheses on the part of the analyst.

In Exploratory Analysis, one aims to find suitable hypotheses to confirm or refute. Automatic discovery automates the process of exploratory data analysis, allowing unskilled analysts to explore very large datasets much more effectively.

The data warehouse may be a suitable place for performing Confirmatory Analysis and looking at the data space. However, it is certainly no place for performing Exploratory Analysis due to the unexpected nature of the queries posed to the data. The natural place for Exploratory Analysis is a data mine and not a data warehouse.

THE PARADOX OF WAREHOUSE PATTERNS

The concepts of *large warehouse* and *useful pattern* often interact in a seemingly paradoxical way. On one hand, the larger a warehouse, the richer its pattern content; for example, as the warehouse grows, the more patterns it includes. If we analyze too large a portion of a warehouse, patterns from different data segments begin to dilute each other, and the number of useful patterns begins to decrease! So, the paradox may be stated as follows: *The more data in the warehouse, the more patterns there are, and the more data we analyze the fewer patterns we find!*

The basic idea is shown in Figure 5.4, but a few simple examples easily clarify this. First, consider a large data warehouse that includes details of a bank's customer accounts, marketing promotions, and so on. Several business objectives may exist for mining this data, including campaign analysis, customer retention, profitability, risk assessment, and so on. To begin with, these are distinct business tasks, and it does not make sense to mix the analyses. Therefore, each of the data mining exercises needs to be performed separately and will require different data structures as well, because some are association analyses; some are clusterings; and so on.

However, even the campaign analysis task itself should often not be performed on the entire warehouse. The bank may have undertaken 30 different marketing campaigns over the years, and these campaigns usually will have involved different products and gone to different customer segments; some of the products are even discontinued now. To understand who responds best to marketing promotions, we need to analyze each campaign (or group of campaigns) separately, because each case will involve patterns with distinct signatures. Mixing the analyses into one data mining exercise will dilute the differences between these signatures. And, the campaigns are

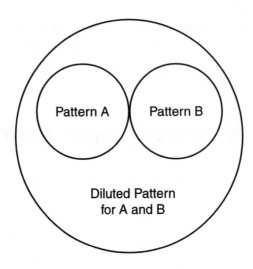

FIGURE 5.4.

Patterns become diluted when you analyze too large a portion of data.

often different enough that mixing them simply may not make sense. So, we need to have a separate Analysis Session for each group of campaigns.

To demonstrate this with a simple example, assume that those customers who are over 40 years old and have more than two children have a high response rate to credit card promotions. Next assume that customers who are younger than 40 years old and have only one child are good prospects for new checking accounts. If we combine these campaigns within the same data mining study and look for customers who have a high response rate, these two patterns will dilute each other.

Of course, we can get a rule that separates these campaigns and still displays the patterns, but in a large warehouse, so many of these rules will appear that they will overwhelm the user. Thus, the smaller patterns may be found in the warehouse if we are prepared to accept large amounts of conditional segment information, for example, 'If Campaign = C12 and ... Then ...'. However, in a large warehouse, so many of these exist that the user will be overloaded with them. The best way is to analyze each group of campaigns separately.

The need for segmentation is even more clear when we consider predictive modeling. When trying to predict the response to a new campaign, it simply does not make sense to base the predictions on all previous campaigns that have ever taken place, but on those campaigns that are most similar to the one being considered. For instance, responses to campaigns for a new checking account may have little bearing on responses to campaigns for a new credit card or refinancing a home. In this case, the paradox of the warehouse patterns comes into play in that by considering more data, we lose accuracy. This is, of course, because some of the data will not be relevant to the task we are considering.

But what happens if one or two key indicators are common to all of the campaigns? Will they be lost if we just analyze the campaigns a few at a time? Of course not. If a pattern holds strongly enough in the entire database, it will also hold in the segments. For instance, if the people with more than five children never respond to campaigns, this fact will also be true in each individual campaign.

As another example, consider a vehicle warranty database. In order to find patterns for customer claims, it is essential to store details of each claim in a large data warehouse. But, does it make sense to analyze all of the warehouse at the same time? No. In practice, cars are built at different plants, and different models of cars use different parts, and some parts are now discontinued. Moreover, over the course of years, the parts used in cars change, so analyzing the entire warehouse may tell us less than analyzing part of it. What works best in practice is to analyze the claims for a given model year for cars built at a given plant – again a segmentation task. The paradox of the warehouse comes into play here in that by analyzing all of the warehouse at once, we reduce the number of useful patterns we are likely to get!

Most of the time, it does not make sense to analyze all of a large warehouse because patterns are lost through dilution. To find useful patterns in a large warehouse, we usually have to select a segment (and not a sample) of data that fits a business objective, prepare it for analysis, and then perform data mining. Looking at all of the data at once often hides the patterns, because the factors that apply to distinct business objectives often dilute each other. The thirst for information can go unquenched by looking at too much data.

THE PITFALLS OF SAMPLING AND SUMMARIZATION

Although I have emphatically used the term "segment" rather than sample several times, it is still worthwhile to explicitly point out the shortcomings of sampling and summarization as potential candidates for anything. Although sampling may seem to offer a shortcut to faster data analysis, the end results are often less than desirable.

Sampling was used within statistics because it was so difficult to have access to an entire population; for example, one could not interview a million people, or one could not have access to a million manufactured components. Therefore, sampling methods were developed to allow us to make some rough calculations about some of the characteristics of the population without access to the entire population. But, does this not fly in the face of having a large database altogether? We build databases of one million customers' behavior in order to have access to the entire population. Otherwise, we could just keep track of a small group of customers.

Segmentation is an inherently different task from sampling. As we segment, we deliberately focus into a subset of the data (for example, one model year for a car or one campaign), sharpening the focus of the analysis. However, when we sample data, we lose information, because we throw away data not knowing what we keep and what we ignore.

The hardware technology for storing and analyzing large datasets provides an unprecedented opportunity for looking at historical patterns by making more data than ever before accessible for analysis. Sometimes, it may seem daunting to look a really large dataset straight in the eyes and try to analyze so much data. It is tempting to try and obtain a smaller sample of the data to build a predictive model. This shyness to look at the whole data is often very expensive, and in most cases, the temptation to sample must be resisted.

At times, when we have a retail database of a million records, it may be suggested that a 100,000 record sample may be good enough. This is not so. Sampling will almost always result in a loss of information, in particular with respect to data fields with a large number of nonnumeric values.

It is easy to see why this is the case. Consider a warehouse of 1,000 products and 500 stores. A half-million combinations exist for how a product sells in each store. However, how one product sells in a store is of little interest compared to how products sell together in each store – a problem known as Market Basket Analysis. For example, how often do potato chips and beer sell together? Five hundred million possible combinations exist here, and a 100,000 record sample can barely manage to scratch the surface. Therefore, the sample will be a really rough representation of the data and will ignore key pieces of information. In using a small sample, one may as well ignore the product column! Hence, we no longer have a large database, because in effect we have reduced it by removing fields. Sampling a large warehouse for analysis almost defeats the purpose of having all the data there in the first place!

Apart from sampling, summarization may be used to reduce data sizes. But, summarization can cause problems, too. In fact, as shown in Figure 5.5, the summarization of the same dataset with two sampling or summarization methods may result in the same result, and the summarization of the same dataset with two methods may produce two different results.

As another intuitive example of how *information loss* and *information distortion* can take place through summarization, consider a retail warehouse where Monday to Friday sales are exceptionally low for some stores, although weekend sales are exceptionally high for others. The summarization of daily sales data to weekly amounts will totally hide the fact that weekdays are money losers, and weekends are moneymakers for some stores. In other words, key pieces of information are often lost through summarization, and there is no way to recover them by further analysis.

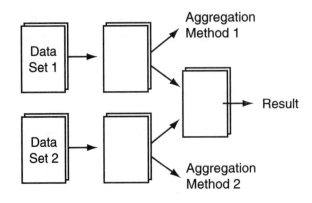

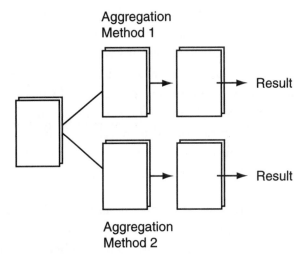

FIGURE 5.5.

Summarization can cause problems.

EXCEPTIONS TO THE RULE

Having established general criteria for where analysis takes place, please note that (as expected) there are exceptions to these rules. Do we ever sample the warehouse for analysis? In some cases, yes. And, do we ever perform analysis on the entire warehouse rather than within a data mine? In a good number of cases, yes.

Sampling is sometimes recommended to get a general feeling for the data, and in such cases, I would recommend having and comparing several samples. Sampling is done only when the computing power is not sufficient to manage the task at hand within a given timeframe.

The question of analysis on the entire database is a slightly different, yet related, matter. In some cases, segmentation alone will not give us all the answers, and we may need to look at some of the overall characteristics of the database. But, even in these cases, we need not perform full exploratory analysis and can compare some of the distributions within the warehouse with those within the data mine.

For instance, as we perform affinity or market basket analysis, we may want to compare how two products sell together in all stores, compared to how they sell together within specific clusters of stores analyzed within a data mine. However, some of these computations may be routinely performed when data is refreshed within the warehouse, and the overall distributions may be kept for such comparisons. In some cases, we may have to perform these computations in a live setting on the warehouse.

And, in some cases, we may wish to perform *segmented affinity analysis* on the entire warehouse in order to find out how to segment the data. Thus, specific operations on the entire warehouse are performed first in order to guide the segmentation process. Therefore, we do sometimes need to look at the warehouse as a whole, but we need to do so selectively and with proper planning.

THE CONCEPT OF AN ANALYSIS SESSION

When using a data mine, we bring a segment (and not a sample) of data from a warehouse (or other sources) to the data mine and perform discovery or prediction. The process of mining this data segment is called an *Analysis Session.* For example, we may want to predict the response to a proposed direct mail campaign by analyzing previous campaigns for similar products, or we may want to know how customer retention has varied over various geographic regions, etc.

An analysis session may be either structured or unstructured. A structured session is a more formal activity in which we set out with a specific task, for example, analyzing profitability by customer segments and/or products. In fact, structured sessions are often performed in a routine manner – we may analyze costs, revenues, or expenses every quarter and understand the reasons for the trends. Or, we may routinely perform forecasting for various items such as product demand in various markets. Or, we may look for unusual transactions that have taken place in the past 30 days. In fact, a structured analysis session usually is of three forms: a *discovery, prediction,* or *forensic analysis* activity where we perform a specific task.

An unstructured session is a wild ride through the database, where the user wanders around without a goal, hoping to uncover something of interest by serendipity, or by help from an exploration agent. This type of abstract wild ride usually uncovers some very wild facts hidden in the data. And, the mine is a natural place for this activity because the unexpected nature of queries may interfere with the more routine tasks for which the warehouse was designed, for example, looking up the history of a specific claim.

The data in the data mine often needs to be enriched with aggregations. Again, let me emphasize that these are not just summaries, but additional elements added to the data. How these aggregations are built is partly decided by a business analysis. For instance, we may need to look at the number of credit cards a customer has as an item. And, we may want to look at the volume of transactions the customer has had. We may also want to look at the number of claims a customer has had in an insurance setting, and so on. These aggregations enrich the data and coexist with the atomic level data in the mine.

THE WAREHOUSE, THE MART, AND THE MINE

The three separate components to an enterprise-wide decisions support system are

● *The Data Warehouse, where the mountain of corporate data is stored all in one place.* Here, data volumes are very high as multiterabyte data warehouses are beginning to appear more frequently. These designs are usually either star-schemas (snowflakes, and so on) or highly normalized data structures.

● *The Data Mart where departmental data is stored and often various external data items are added.* The data volumes are usually 15 percent to 30 percent of warehouse sizes, and the envelope is being pushed toward the terabyte limit. These databases are also usually either based on star-schemas or are in a normalized form. They mostly deal with the data space, but sometimes multidimensional analysis is performed.

● *The Data Mine, where the data is reorganized for analysis and information is extracted from the data.* The data volumes here are the same as the Data Mart, but the data is much more richly structured and is no longer just departmental. The data here refers to a specific business objective and is analyzed for the purpose of information extraction.

Although the data structures used within the warehouse and the data mart may be similar, the data structures used within the data mine are significantly different. The data mine differs from the data warehouse not just in the size of data it manages but in the structure of the data. The content of the data in the mine is also often different from the data in the warehouse, because it is often enriched by additional external data not found within the warehouse. However, content aside, the key issue about data mining architecture is that the existing theories of data structuring do not apply to it.

The two key approaches to data structuring within warehouses are *normalization* and *star-schema families*, which include snowflake schemas. When we think of the origins of these approaches and the four spaces of decision support, it is not surprising that these two design methodologies do not easily lend themselves to successful use within a data mine.

Normalization theory was invented by Ted Cod in the 1970s as a theory and methodology for data structuring for OLTP applications, and it produced extremely good results in the 1980s, during the rush toward the deployment of relational databases for operational purposes. In fact, without normalization theory, many of the successful large-scale database projects of the 1980s and early 1990s would have failed altogether.

However, the limitations of normalization theory became evident when it was applied to dimensional analysis for decision support. The observations made by Ralph Kimball and others about the dimensional data needs of large retailers gave rise to the introduction of star-schemas and database engines, such as Redbrick, which successfully took advantage of these ideas in the late 1980s. Although normalization theory deals with the data space, star-schemas deal with the aggregation space. By the same token, although star-schemas may be suitable for dimensional analysis in the aggregation space, they are not ideal for data mining within the influence space, because the structure of this space is logical and not arithmetic or polynomial.

The data structures in the data mine need to be both denormalized and super-dimensional. The details of a new theory of structuring for data mines based on Rotational Schemas is beyond the scope of this book and will be presented elsewhere. However, note that within the data mine, we sometimes need to look at super-dimensions that exist above the dimensions used in the OLAP space, for example, the data mining dimensions subsume the OLAP dimensions.

CHAPTER 6

Data Mining Above, Beside, and Within the Warehouse

Topics in this chapter:

- Stand-alone Data Mines

After we accept the fact that the data mine is distinct from the data warehouse, the next logical question is: Where does the data mine actually exist? Is it a separate repository next to the warehouse, a set of views above the warehouse, or just part of the warehouse? We can answer this question in each of these three ways and get a different architecture for the data mine.

The data mine can exist in three basic forms:

- *Above* the warehouse, as a set of conceptual views

- *Beside* the warehouse, as a separate repository

- *Within* the warehouse, as a distinct set of resources

Data mining "above the warehouse" provides a minimal architecture for the discovery and analysis. It is suitable only in cases in which data mining is not a key objective for the warehouse. In this approach, as shown in Figure 6.1, SQL statements are used to build a set of conceptual views above the warehouse tables. And, additional external data from other tables may be merged as part of the views.

The views built above the warehouse may either be materialized (for example, saved to disk as new tables) or not. Therein lies the fundamental problem (if not contradiction) built into this approach. If the views are not of significant size, serious data mining cannot take place. However, if the views are of a significant size, without materialization, the effort in computing them again and again will require very large amounts of processing power – in some cases significantly affecting the availability of the warehouse resources and interfering with other applications performing indexed retrievals.

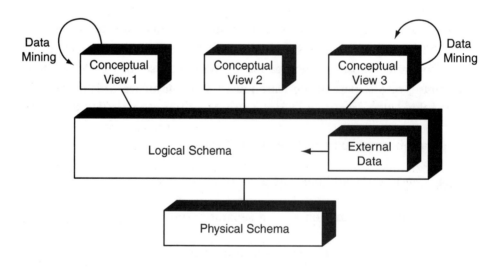

FIGURE 6.1.

Data mining above the warehouse.

If the views are of significant size and they are materialized, we are no longer data mining above the warehouse and will be using a disorganized form of the third approach, for example, data mining within the warehouse. If the views are materialized, the third method will almost always work better, because it can use a suitable data distribution approach and a specific processor allocation strategy, as well as different data structures for data mining. Without these precautions, the number of potential pitfalls increases rapidly, sacrificing both performance and functionality.

Therefore, data mining above the warehouse should be restricted to applications in which data mining is only of peripheral business interest and not a key objective. However, holding this view is often a big business mistake in itself; why have so much data in a warehouse and not understand it?

In most cases, data mining is effectively performed beside the warehouse, with data structures that lend themselves to detailed analyses. And, in most cases, additional data suitable for the analyses is merged with the warehoused data in order to perform specific analyses for focused business needs.

The concept of data mining beside the warehouse fits well within the context of three-level computing, which resembles a three-tiered client server architecture. Within a three-tiered, client-server architecture, a first-tier client interacts with a middle-tier server that also interacts with a third-tier large system. Although in three-tiered, client-server systems the interaction between layers 2 and 3 is ongoing, in a three-level computing system, most of the interaction is between the client and a specialized server that occasionally accesses a huge server. In other words, the huge server holds very large amounts of data and is surrounded by a number of specialized servers, which interact with the clients.

The overall architecture for data mining beside the warehouse is shown in Figure 6.2. Here, the process of data migration and fusion populates a data warehouse with large amounts of historical data. The data structures used within the warehouse may be either normalized or members of the star-schema family. However, the data schemas for the data mine will be different.

Query and reporting tools, as well as other applications using traditional database index structures may directly access the warehouse with very good results; the warehouse will support access to the data space. However, detailed analyses, such as discovery and prediction, are not performed within the warehouse because they do not relate to the data space. Instead, these activities are performed in the data mine, with data structures suitable for data mining.

Data mining beside the warehouse both overcomes and sidesteps several problems at once. It allows data mining to be done with the right data structures, avoiding the problems associated with the structures of the data space. Moreover, the paradox of warehouse patterns can be avoided by selecting specific data segments, corresponding to specific business objectives. Also, the interactive exploratory analyses that are often performed in the data mine with wild rides through the data no longer interfere with the warehouse resources that are responsible for routine processes, such as query and reporting.

In fact, different business departments can use their own data mines that address their specific needs; for example, direct marketing versus claim analysis. The data is then moved from the large warehouse to the mine, is restructured during the transformation, and is analyzed. It is, however, important to design the transfer and transformation methods carefully, in order to allow for optimal refresh methods that require minimal computing. For instance, as we bring new data into the data mine every day or every week, the overhead for reaggregation should be minimized.

In some cases, where the warehouse is a very large, massively parallel processor (MPP), the data mine may actually reside as a separate repository within the large warehouse. As shown in Figure 6.3, this is very similar to a data mine beside the warehouse, where the mine uses a portion of the physical warehouse but is independent of the warehouse structures, in effect being a "republic within a republic."

In this scenario, the disk apace and the processors for the data mine are specifically allocated and separately managed. For instance, on a shared

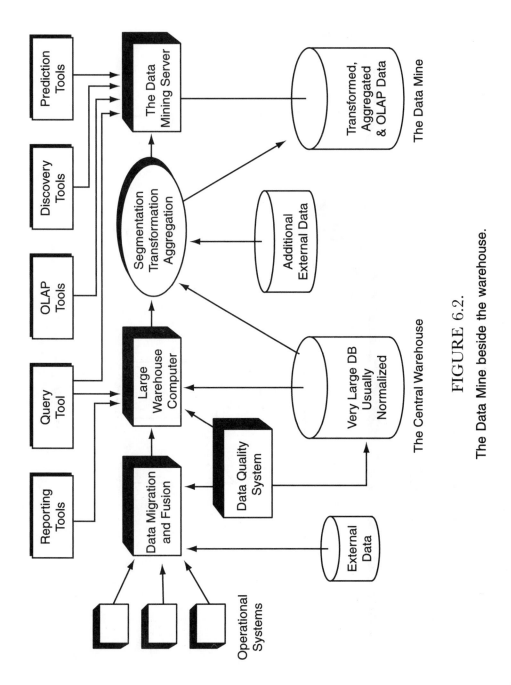

FIGURE 6.2.

The Data Mine beside the warehouse.

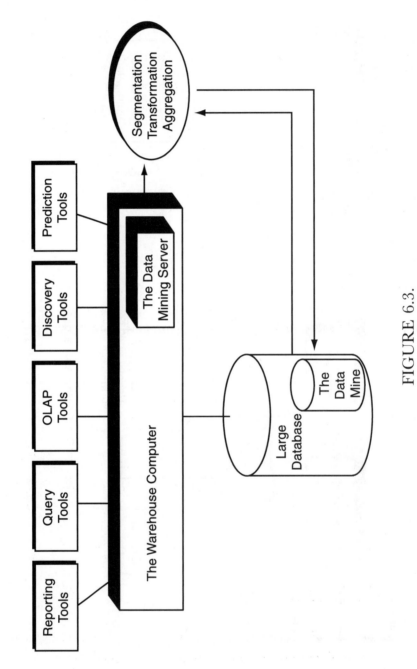

FIGURE 6.3.

The Data Mine within the warehouse.

nothing MPP machine with a 32 processor, the disk space for the data mine is separately allocated on eight of the 32 nodes, and eight processors are dedicated to data mining, while the other 24 processors manage the rest of the warehouse. And, when needed, additional processing capability may be directed toward the needs of data mining.

Although this idea may sound attractive based on arguments for centralization and scalability, in practice it usually leads to loss of flexibility, without providing any significant benefits for data mining. In most cases, when we consider the technical, marketing, and business issues, the problems with mining within the warehouse multiply quite rapidly, and the data mines planned for use within the warehouse will eventually find themselves beside it.

The key point is that the likelihood of serving the needs of many people within the data space is much higher than the likelihood of serving their needs within the multidimensional and influence spaces. Although the data elements may be almost the same for several departments, the dimensions, the influence relationships, and the predictive models they all need will vary far more than their simple data needs. Therefore, the data mine within the warehouse will soon become the lowest common denominator of all designs.

Although the design of the data space may be subject to compromises to please the various user groups, no compromises should be made in the design of the data mine where serious and detailed analyses take place. The data mine should be optimized to deliver effective results by focusing on specific business needs, because influence analysis is so much harder than data access.

STAND-ALONE DATA MINES

Can a data mine exist without a warehouse? In some cases, yes. This usually happens when a business unit within an organization needs to get results next quarter, not next year. When it takes too long to build the very large corporate warehouse, some business units have no choice but to take things into their own hands – often with excellent results!

And, coupled with the Sandwich Paradigm, a stand-alone data mine is the best way to guide the data warehouse design process. Often, a significant portion of the business benefits from the system are to be obtained from the data mine, and the horizontal and vertical prototypes built with the Sandwich Paradigm will make sure that the eventual system will conform to the implicit needs of the user.

The key issue of concern for managing a stand-alone data mine is data fusion and quality management. However, these are problems that a large warehouse faces anyway, and dealing with them in the context of a data mine is not much more difficult, although they still do need to be handled with a great deal of care and attention.

The need for stand-alone data mines usually arises when large warehousing projects have to deal with so much data and need to please so many different departments with distinct requirements that they are inevitably forced into becoming a "QE2 Sized Solution;" they attempt to build the largest ship possible, hoping to carry all passengers to any port. Yet, it often takes too long to deliver the QE2, and it is later discovered that a ship that size cannot sail to every port for data analysis. In such cases, stand-alone data mines become a very attractive option, delivering rapid and focused benefits.

Stand-alone OLAP engines have already followed this very path to success. Many a departmental system for multidimensional analysis has been effectively deployed while the meeting dates for the design of the corporate data warehouse were being discussed. Thus, stand-alone data mines deployed prior to the warehouse not only provide a Sandwich Paradigm effect by clarifying the needs of user groups, but deliver significant business benefits during the two years it takes to deliver a QE2-sized data warehouse.

CHAPTER 7

What Is Metadata?

Topics in this chapter:

- Abstraction

- Use of Metadata in the Data Warehouse

The average reaction to any mention of metadata is probably, "This conversation has suddenly become rather technical. Is this really the time? I must be off!" Despite its discouraging name, though, the subject of metadata is quite easy to understand. Understanding metadata is essential for the success of any serious information-handling project, such as a data warehouse.

The standard definition of metadata is *data about data*. Although it serves as a handy reminder, this definition is hardly very illuminating. For a start, how can we add meaning to data by associating it with more data?

ABSTRACTION

Alfred Korzybski, founder of the discipline of General Semantics, introduced the idea of a *chain of abstraction*. For instance, an actual chair is a pretty concrete item. Any specific chair has a host of characteristics: It is made of wood; it has two arms; it is a particular shape; it has two cushions;

and so on. We can abstract from that chair, and others we encounter, to the word "chair." That is one step in the chain of abstraction, but, of course, we can take more steps. For instance, we could describe the word "chair" and all the other words we know by the word "word." There is scarcely any limit: "word," "ideal," "motive," and "hatred" are all instances of "abstraction," and so on.

Data and Metadata

Software is abstraction piled on abstraction. The word "chair" that you are reading now, if stored in a computer, would be represented by a string of flux changes on the oxide surface of a hard disk. As would a source code statement like the following:

```
"LET BALANCE = BALANCE + CREDIT".
```

The source code statement, in turn, is an abstraction of a banking transaction.

Metadata is an abstraction from data. It is high-level data that describes lower-level data. Software is full of metadata, for example:

- Record descriptions in a COBOL program

- Logical view descriptions in a data server's catalog

- SQL Create statements

- Entity-relationship diagrams in a CASE tool's repository

Metadata Is the Key to Data

Metadata is instrumental in transforming raw data into knowledge. For instance, metadata in the shape of a field definition tells us that a given stream of bits is a customer's address, part of a photographic image, or a code fragment in a given computer's machine language. Should the metadata get mixed up or out of alignment, none of the data will make sense. The address will contain nonsense characters; the code will give unpredictable results and probably crash any process that runs it; even the photograph will not look quite the way it should, although our eyes may not detect the difference.

Suppose that a large multinational corporation is building a data warehouse and loading it with data and metadata from departmental and local databases. How likely is it that all the different sets of data and metadata will agree? In one such case, more than a dozen different ways existed to encode the unique badge number allocated to each employee. If the French subsidiary uses a 32-bit integer for the purpose and the Australians use binary coded decimal, what are the chances that applications written to use these two representations could interpret each other's data? Organizations are increasingly turning to external sources of information, both in search of impartial data and to draw on the expertise of specialists in other fields of knowledge. Still more scope for confusion and misunderstanding is possible in the interpretation of external data. Indeed, information from one source may flatly contradict another.

It may be helpful to think of metadata as "tongs" with which we can handle raw data. Without metadata, the data is meaningless. We do not even know where it is or how much of it to take. Take away data definitions, and how could we query a database? All we could hope to recover would be a string of 1s and 0s. In the early days of commercial computing, each application created and handled its own data files. Because the files' metadata was embedded in the application's data definitions, no application could make sense of another application's files. When databases were introduced, one of their greatest advantages was that the metadata was stored in the database catalog, not in the individual programs. There was one version of the truth, and that version was held in the database.

When database manuals talk about *structured data*, they are implying the presence of metadata. Relational databases can store information for which they have no corresponding predefined data types, but this is not a pretty sight. *Binary large objects* (BLOBs) are a case in point. BLOBs are used to store large complex data items, such as maps, medical X-rays or scans, photographs, or CAD diagrams. Because relational databases lack the metadata to describe these items, the relational vendors call them unstructured data. However object-oriented databases can store these items and many others in a structured format, even including the code routines needed to make sense of them.

How is metadata used in the data warehouse?

USE OF METADATA IN THE DATA WAREHOUSE

Metadata in Action

Consider for a moment a business analyst rummaging through a data warehouse and finding three sets of data:

- 739516 13238350 426615 800441 4912313

- ¢An Inverness Group report dated 4/7/95 states that the European market for repository tools expanded by 33% in 1994¢

- ¢Leading gadget vendors: Protz Group 48%, Harris Goods 29%, Zymurgy Inc. 13%¢

How much can the analyst safely deduce from this information?

In the first case, it is fair to say that the answer is "absolutely nothing." The numbers listed may be a firm's sales for different regions, population of towns, or the number of hairs on certain individuals' heads. The numbers might even represent a sequence of machine code for some computer. This is what one would normally expect of tabular data from a database. The only ways to assign any meaning to it are

- *From the context* – If this data is the result of a query on a given table, we already know its meaning.

- *From metadata* – If we can associate a metadata description with the data, it will tell us the name of the table and possibly a great deal more.

The second example appears more straightforward. It is free text and seems to be self-describing. One small point, however: The date is ambiguous. Does it mean "4th July 1995" (British convention) or "7th April 1995" (U.S. convention)?

The third example contains some metadata, but not enough. Stipulating that "gadget" has a precise meaning to someone in the gadget industry, we still do not know whether the market referred to is United States, world, European, or other; what time period is referred to; how the data was collected; or even the source of the information. Under the circumstances, this example should be filed in the data warehouse with a metadata note that says "source unknown, reliability unknown, no further details."

Metadata in the Operational and Data Warehouse Environments

In the operational environment, metadata is mostly valuable to software developers and database administrators. Operational databases are accessed only by transaction processing applications, which contain data definitions embedded within them. The end users, people like bank clerks, travel agents, and hospital staff, do not need to know how information is held in the database. They interact with the forms and screens provided by the applications they use in the course of their work.

The decision-support environment is very different. Here, data analysts and executives are looking for useful facts and correlations that they will recognize when they find them – and often, not before. Routine applications are of no use to them; they need to get in among the data, and to do so successfully, they need to understand its structure and meaning. Passengers on a train do not need a map, although a timetable may prove useful. A driver setting out by road across a foreign country to an unknown village, however, would be very unwise to set out without a complete set of large-scale and small-scale maps. As Bill Inmon of Prism Solutions, inventor of the data warehouse, likes to say, a family that moves into a new house usually pays careful attention to the local road signs for the first week or so. Then, during the following years, they hardly notice them at all. This is not because the road signs do not matter, but because the drivers have come to know them by heart.

A data warehouse without adequate metadata is like a filing cabinet stuffed with papers but without any folders or labels. Try finding anything in that!

Generally speaking, three main layers of metadata exist in a data warehouse:

- *Application-level (or operational) metadata* – This metadata defines the structure of the data held in operational databases and used by operational applications. Application-level metadata is often quite complex and tends to be application- or department-oriented.

- *Core warehouse metadata* – This metadata is held in the catalog of the warehouse's database system. According to Inmon, it is distinguished by being subject-oriented – in other words, it is based on abstractions of real-world entities like "customer," "project," or "organization." Core warehouse metadata defines the way in which the transformed data is to be interpreted, as well as any additional views that may have been created, including decision-support aggregates and computed fields, as well as cross-subject warehouse views and definitions.

- *User-level metadata* – This metadata maps the core warehouse metadata to business concepts that are familiar and useful to end users.

What Metadata Is Required?

Avoiding the Credibility Gap — the Need for Consistency

Unreliable or missing metadata leads to the familiar situation in which one department tells the CEO that corporate profits are up 10 percent, and another department says they are down 15 percent. Each department is using its own figures, collected in accordance with its own procedures and interpreted by its own applications. The discrepancies produced in this way are sometimes referred to as the *algorithmic differential.* In the absence of hard facts, decision making lapses into a morass of politics and personalities.

Most corporations have tried to implement a data dictionary of some kind – some even got involved in IBM's disappointing AD/Cycle initiative – and came to consider the whole technology a failure. Developers and data administrators were supposed to keep the dictionary up to date at all times, but they saw this as an extra chore that stole time from their real work. After a few changes have been made to a system without corresponding updates to the dictionary, that dictionary is dead meat. It is no longer synchronized with the live system, and everyone can legitimately ignore it. From then on, whenever metadata is needed, people get it in the same way they always have done – by asking the oldest data administrator (until he or she retires).

The data warehouse, however, is set up for the benefit of business analysts and executives, and the people who create and administer the warehouse are there to help the end users get their work done. If they need metadata – and they do – it is the job of the warehouse support staff to make sure that they get it. A poor head librarian would allow library users to wander around at random looking at the spines of books until they happened on the title they wanted. Instead, every library has an index on cards, microfiche, or computer that indexes books by author, subject, and title. For those whose requirements are less precise, libraries usually divide their books up into rather coarse sections, such as Fiction,

Science, Sports, and so on. Under Sports is a shelf on tennis, a shelf on football, and so on.

Planning for Change

In the data warehouse, not just the data but the metadata keeps on changing. The business users are continually looking for interesting new patterns, and this can lead them to compare information from the four corners of the enterprise. This quest also can lead them to keep asking for new sets of data that have to be replicated from operational systems or imported from outside sources. The warehouse's metadata map has to be extended to embrace each new addition. Besides, many business entities, such as product lines, organizational structures, markets, and plans, change regularly. Inmon cites the example of a business analyst asked to prepare a report for the CEO. The analyst works hard and delivers the report, complete with impressive graphics, within 24 hours. The CEO congratulates him and asks for the same information on the period five years ago. After a lot more work, this, too, is forthcoming, but the CEO is displeased. After a brief glance at the two reports, he complains that the IS department can never get anything right. These figures are obviously wrong. How can this be? According to Inmon, some of the factors to evaluate are

- Different sources of data

- Different sales territories (perhaps with the same names)

- A different definition of the product line

- Tax has been added in, or removed, from the figures

- Mergers leading to sudden jumps in revenue, profit, and many other metrics

After these factors have been brought to his notice, a reasonable CEO would admit that the analyst had done another good job. He would never have thought otherwise, if the various changes in the basis of measurement been noted on the report.

A Single Version of the Truth

Not only must warehouse administrators keep up with ever-changing corporate definitions, they must also agree on a single version of the truth. An executive in one multinational corporation was heard to remark that, if the revenues for which credit was claimed by all the product line managers were totaled, the resulting value would exceed the corporation's actual revenue by a factor of five. In another company, a marketing manager found that the products for which he was responsible were 15 percent under budget for the quarter. A strenuous midnight consultation of various databases revealed that the manager of a systems integration unit had unilaterally created a set of new part numbers, under which were subsumed all the products sold as part of a given deal. This innovation, which greatly simplified life for the corporation's customers, had the unfortunate side effect of diverting several million dollars worth of products from the marketing manager's top line and adding them to the systems integration manager's. After these facts were ascertained, it took a matter of minutes to set things to rights. But, only a manager who was on intimate terms with metadata would have been able to uncover the facts.

Inmon identifies at least two contrasting kinds of metadata:

- *Classical* – This metadata consists of formal definitions, such as a COBOL layout or a database schema.

- *Mushy or Business* – This metadata consists of information in the enterprise that is not in classical form – for example, the organization chart or historical pricing information. (Most businesses do not formally retain pricing information from previous years.) As Inmon puts it, "this stuff is in people's heads or on paper." To look at it another way, there is no meta^{2}data (metametadata) for it, no formally prescribed process for recording the metadata.

Metadata Requirements

According to Inmon, a new analyst approaching a data warehouse wants to know

- What tables, attributes, and keys does the data warehouse contain?

- From where did each set of data come?

- What transformation logic was applied in loading the data?

- How has the metadata changed over time?

- What aliases exist, and how are they related to each other?

- What are the cross-references between technical and business terms? (For instance, the field name XVT-351J presumably meant something to a COBOL programmer in 1965, but what does it mean today?)

- How often does the data get reloaded?

- How much data is there? This helps end-users to avoid submitting unrealistic queries. Given some means of determining the size of tables, staff can tell the end users, "You can do anything you like with 15,000 rows, but if it turns out to be 15 million rows, back off and ask for help!"

Metadata Components

Warehouse metadata is not very different in kind from ordinary database metadata, although it is versioned in order to permit historical analysis. Prism gives the following breakdown of warehouse metadata in its Tech Topic, "Metadata in the Data Warehouse:"

Mapping

The mapping information records how data from operational sources is transformed on its way into the warehouse. Typical contents are

- Identification of source fields

- Simple attribute-to-attribute mapping

- Attribute conversions

- Physical characteristic conversions

- Encoding/reference table conversions

- Naming changes

- Key changes

- Defaults

- Logic to choose from among multiple sources

- Algorithmic changes

Extract History

Whenever historical information is analyzed, meticulous update records have to be kept. The metadata history is a good place to start any time-based report, because the analyst has to know when the rules changed in order to apply the right rules to the right data. If sales territories were remapped in 1991, results from before that date may not be directly comparable with more recent results.

Miscellaneous

- Aliases can make the warehouse much more use-friendly by allowing a table to be queried by "Widgets produced by each factory" rather than "MF-STATS." Aliases also come in useful when different departments want to use their own names to refer to the same underlying data. Obviously, though, aliases can also cause a great deal of confusion if they are not carefully tracked.

- Often, parts of the same data warehouse may be in different stages of development. Status information can be used to keep track of this: for instance, tables might be classified "in-design," "in-test," inactive," or "active."

- Volumetric information lets users know how much data they are dealing with, so that they can have some idea how much their queries will cost in terms of time and resources. Volumetrics could usefully include such information as number of rows, growth rate, usage characteristics, indexing, and byte specifications.

- It is also useful to publish the criteria and time scales for purging old data.

Summarization Algorithms

A typical data warehouse contains lightly and heavily summarized data as well as full detailed records. The algorithms for summarizing the detail data are obviously of interest to anyone who takes responsibility for interpreting the meaning of the summaries. This metadata can also save time by making it easier to decide which level of summarization is most appropriate for a given purpose.

Relationship Artifacts and History

Data warehouses implement relationships in a different way from production databases. Metadata pertaining to related tables, constraints, and cardinality will be maintained, together with text descriptions and ownership records. This information and the history of changes to it can be useful to analysts.

Ownership/Stewardship

Operational databases are often owned by particular departments or business groups. In the nature of an enterprise data warehouse, however, all data is stored in a common format and accessible to all. This makes it necessary to identify the originator of each set of data, so that inquiries and corrections can be made to the proper group. It is useful to distinguish between *ownership* of data in the operational environment and *stewardship* in the data warehouse.

Access Patterns

It is desirable to record patterns of access to the warehouse in order to optimize and tune performance. Less frequently used data can be migrated to cheaper storage media, and various methods can be used to accelerate access to the data that is most in demand. Most databases do a good job of hiding such physical details, but specialized performance analysis tools are usually available. Some general-purpose tools, such as Information Builders' SiteAnalyzer, also are available.

Reference Tables/Encoded Data

Reference data is stored in an external table and contains commonly used translations of values. The contents of these tables must be stored in order to guarantee the ability to recover the original unencoded data, together with effective from and effective to dates.

Data Model — Design Reference

Building a data warehouse without first constructing a data model is very difficult and frustrating. When a data model is used, metadata describing the mapping between the data model and the physical design should be stored. This allows all ambiguities or uncertainties to be resolved.

CHAPTER 8

Tools for Metadata

Topics in this chapter:

Many well-known software tools have long been used for generating metadata in other contexts. These tools include CASE tools, and, to a limited extent, compilers and 4GLs. More recently, vendors like Prism, Carleton, and ETI have designed products specifically to create data warehouse metadata. Specialists like BrownStone, Reltech (both of whom have now been acquired by Platinum Technology), and Rochade have been declaring that their repositories can play a valuable role in the warehouse environment. These repositories have ample power reserves to store metadata from the underlying operational systems, the business rules used for transformation, and end-user views. Repository browsers, such as the Reltech Data Shopper, IBM's Data Guide, and IBI's EDA/SAF, help administrators and users to browse a repository's structure and explore its resources.

According to Bill Inmon, conventional CASE tools are of limited value, as they assume a sequential waterfall lifecycle; whereas data warehouse construction really demands a RAD approach. However, data modeling tools like Logic Works' ERwin and Powersoft's S-Designor have their places.

Turning to metadata consumers, the first obvious category of tools consists of data transfer tools that load data from operational systems into the warehouse. Many of these tools exist; some of the best known are Prism, Carleton, ETI, Infopump, and IBI's Copy Manager. Different tools store metadata in their own ways. The simplest way to store the metadata is to keep it in the data warehouse's own catalog, but some transformation products keep metadata in a separate database while they are working on it.

Tools, such as Focus and Software AG's Esperant, provide a very useful extra layer of metadata between the end user and the raw database catalogs (or their equally incomprehensible SQL representations). The underlying truth is that a data warehouse is only as good as its facilities for handling metadata. Every time data moves from one place to another, it is vital that the data is correctly and appropriately labeled.

Essential activities, such as filtering, cleaning, summarizing, and consolidation, all involve mapping from one metadata representation to another. For setting up a fast and reliable system of bulk replication from the operational systems to the data warehouse, metadata checking and mapping must be carried out automatically.

A multidimensional model, or star-schema, is the standard technique for delivering data to end users in a data warehouse environment. The design consists of a fact table containing numeric and additive measures as well as foreign keys that link that fact row to a number of separate dimension tables. The dimension tables contain textual or other descriptive items, usually serving as constraints on queries. Items brought together in a single dimension are meant, as a group, to describe some relevant business category to end users. This grouping of items into business categories naturally contains hierarchies of information. Each row in a dimension table has a one-to-many relationship to the fact table. Relatively, the fact table contains fewer columns than the dimension table, but, at the same time, it has many, many more rows than the dimension table. Figure 8.1 shows an example of a generic star-schema.

Using the star-schema for data delivery to end users has many advantages. The design is much simpler than most relational/third-normal form designs. Star-schemas relate well to what users can conceptualize in terms of spreadsheets and even as regression analysis formulae. Typically, users want the data they access to be in a single universe table containing everything they desire, and the star-schema serves as a slightly normalized compromise to accomplishing that need.

The standard paradigm of maintenance for sets of star-schema tables requires that new entries to the fact table be simple inserts of additional transactions at a chosen level of detail (properly called *granularity*). Although changes for the dimension table may be directly applied, those changes must be handled in a particular manner. The compelling reason for this distinction in approach is that changes to the dimension's data must maintain an unaltered relationship between the fact and any given dimension table row. If this relationship is altered, updates to the fact table are necessary to correct these relationships. Due to the great number of rows that can exist in the fact table, applying updates to the fact table can be problematic.

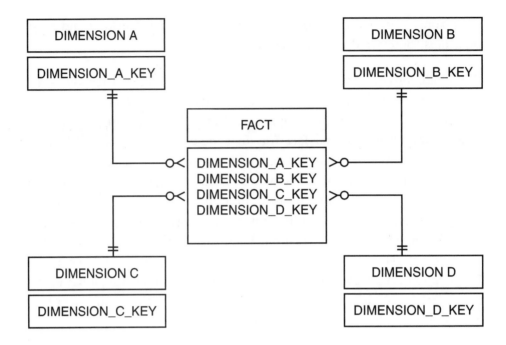

FIGURE 8.1.

A generic star-schema.

In his seminal books on multidimensional design, Ralph Kimball proffered the concept of "slowly changing dimensions" to handle change within a dimension table and defined three approaches to implementation:

TYPE 1. Overwrite the old values in the dimension record with the new values, thereby losing the capability to track the old history.

TYPE 2. Create an additional dimension record using a new value of the surrogate key, thereby segmenting history between the old value and the new value.

TYPE 3. Create an old field in the dimension record to store the immediate previous attribute value, thereby being able to describe history both forward and backward from the change.

Following is a suggestion for revising the perspective on slowly changing dimensions. The intention of this redefinition is to focus on the result rather than the specific implementation. This altered perspective is stated as follows:

TYPE 1. A fact is associated with only the current value of a dimension column.

TYPE 2. A fact is associated with only the original value of a dimension column.

TYPE 3. A fact is associated with both the original value and the current value of a dimension column.

This revised view of slowly changing dimensions promotes a fourth domain:

TYPE 4. A fact is associated with the current value and the values that were current at some critical points in time (for example, end of year, each year) of a dimension column.

Under these altered definitions, *how* we accomplish the fact-to-dimension associations is no longer the sole focus of the various types. The reason for offering this altered view is that people are creative. If we leave the type tied to the specific implementation, we then need to expand our type list every time someone uses his or her creativity. Besides, under the scenario of that ever-expanding type list, who's going to maintain the official list so we don't end up with 10 variations that are implemented by different creative individuals around the world, all of them designating their own version "Type 8?"

For example, suppose that a dimension table is refreshed every cycle. Our processes may be so defined that we retain the integrity of the sequence number keys (or cheat and use the operational data values themselves as the key, as we are *not* supposed to be doing). With sequence number integrity retained, we basically drop, rebuild, and reload our dimension table every cycle. We have not overwritten anything, but from time period to time period our dimension data is indeed changing. Our facts are always associated with the current values of dimension columns. Nominally, this variation is not any one of the original three slowly changing dimension types. The closest association under the original type definitions would be to say, "Conceptually, it falls under the Type 1 umbrella, even though we aren't actually updating." Under the new definition, it is clearly, totally, and simply a Type 1 slowly changing dimension.

More creativity exists. In some implementations, dimension tables have been established with two-part keys: a sequence number identifying an occurrence and a secondary sequence number identifying a version. The fact table contains both key parts. Joins using both parts provide the original value. Joins using just the first part and secondarily adding in a current version indicator on the dimension table provide the original dimension value. In this instance, new values cause the creation of new rows. Under the original type definitions, it means that we have a Type 2 change. However, for this example, processing does not allow for a new unintelligent sequence number key to be generated. Processing must determine which things really are new things and which are new versions of old things, resulting in a rather complicated process that allows for far greater opportunities to make mistakes along the way. This may also require extra lookups, necessitating more computer resources to process a change. Even though user querying must decide on one of two forms for joining to a dimension, a fact can be associated with both a current and with an original value (as shown in Figure 8.2). Consequently, the example scenario actuality describes a Type 3 slowly changing dimension in Type 2's clothing. If we say the change is Type 2 under the original definition, the implication is that only the original value is available in fact analysis, which is not true. If we say it is Type 3 under the original definition, it implies that we have distinct original and current columns in our individual dimension table rows, which is also not true.

Type 4, as previously defined, is quite similar in spirit to Type 3. The sole distinction is that Type 3 allows for only a current and original value to be associated with a fact. Type 4 allows for an infinite number of additional time-valued slices of dimension values to be retained. Because of this addi-

Fact

Dimension Key	Version Number	Measure
001	01	1000
001	01	900
001	01	1100
001	01	800
001	02	900
003	01	1200
004	01	1500
004	01	1100
005	01	900

Dimension

Dimension Key	Version Number	Description	Current Ind.
001	01	ABC	Y
002	01	GHI	N
002	02	JKL	Y
003	01	MNO	Y
004	01	QRS	Y
005	01	WXY	Y

Join 1: Fact. Dimension Key = Dimension. Dimension Key
AND Fact. Version Number = Dimension. Version Number

Fact

Dimension Key	Version Number	Measure
001	01	1000
001	01	900
001	01	1100
001	01	800
001	02	900
003	01	1200
004	01	1500
004	01	1100
005	01	900

Dimension

Dimension Key	Version Number	Description	Current Ind.
001	01	ABC	Y
002	01	GHI	N
002	02	JKL	Y
003	01	MNO	Y
004	01	QRS	Y
005	01	WXY	Y

Join 2: Fact. Dimension Key = Dimension. Dimension Key
AND Dimension. Current Ind. = "Y"

FIGURE 8.2.

A fact can be associated with both an original and a current value.

tional multiplicity, it is certainly worthy of its own recognition. Recently, Kimball has acknowledged that such circumstances can arise and has referred to this as "many alternate realities" with the implementation being a hybrid Type 2/3. Hopefully, the business reasons for doing this will be rare, but if a true reason does exist, it is incumbent on a data warehouse designer to attempt to satisfy that need. Under the original definition of Type 3, Kimball made efforts to explain that indeed such a column addition, to hold both an original and current value for an item, could be done only once. In reality, no DBMS restriction prevents the addition of columns (unless, of course, you have reached your DBMS's physical column limit.) Certainly, the concern here is to do such a thing under only very limited and extreme circumstances. You do not want to have unbounded column proliferation across your star-schema. For example, suppose that we have a star-schema containing 25 years of data (yes, some future time where things are really wonderful). For our business customer information, a critical need exists for retaining their standard industrial classification (SIC) codes over time. The government reworks SIC codes every 10 years for classifying types of businesses. We have columns on our customer dimension table containing a 2000 SIC code, a 2010 SIC code, and a 2020 SIC code. Maybe we are even gearing up to add in a 2030 SIC code. As long as this column proliferation is approached judiciously, this approach to retaining historical change can be functional.

Remember the new type definitions extrapolated in this discussion are concerned with the functionality, not the detailed implementation, of the change protocol. Let's look back at our last Type 3 example. If the second part of our two-part dimension key were a date, instead of a version sequence number, it would be possible for our fact to join to a dimension using the first part of the key and applying a range test against the second part of the key to retrieve the dimension's value at any given point in time (assuming, of course, a DBMS optimizer is capable of effectively performing such a process). Likewise, imagine taking a snapshot of one's dimension at a given point in time – for example, year-end of every year – and retaining that snapshot as a separate table. Depending on the users' analysis needs, users could join to a specific year's version of the dimension table to gather the dimension's values at any of the times for which the snapshot versions exist. In each of these cases, a fact can be associated with dimension values from multiple points in time. These examples would both demonstrate implementations of a Type 4 slowly changing dimension.

It is a column-by-column decision within a dimension table as to which items are to be handled via a Type 1, 2, 3, or 4 approach. To make this work, the combining of the processing needs has to reach the point where complexity exists in maintaining your dimension tables. A star-schema, or multidimensional design, has become an important concept in the data warehousing world. Much of the power of the star-schema is in its consistency of form. In a unique way, the star-schema is its own normalized design. Likewise, language is important. The words we use do much to color our perceptions. It can be argued that, logically speaking, the new wording offered here does not really change anything. The intention is not to change the basic premises; the intention is to offer a refinement in detail. We have no need to insert another semantic layer that designers need to interpret. (A layer requiring designers to identify their implemented procedures as one type of slowly changing dimension or another based on which type can be abstracted to match what they are doing rather than which type matches what they are doing.) Perhaps this refinement can expand our perspectives on examining and implementing slowly changing dimensions. The level of standardization offered within a star-schema still has much that can be written to formalize the design process beyond what exists today. Part of Kimball's goal in framing slowly changing dimensions surely is to prevent a proliferation of "creativity" while guiding efforts to a best-practice approach. Although his concerns are laudable, implementations show that greater variations have been put into practice and do function for organizations. We do not need to exclude these variations from explicit and legitimate coverage under the umbrella of handling change and maintenance within a star-schema via slowly changing dimensions.

2001 TRENDS FOR CRM-READY DATA WAREHOUSES

We are often asked what trends we see for data warehousing and CRM. Although there are quite a number of them, here are listed a few that will undoubtedly receive serious attention this year – the ones related to preparing and expanding data warehouses to support CRM. If your company is following a strategy of competitive advantage through customer relationships, you will want to be sure your CRM-ready data warehouse is enabled to support all activities.

A CRM-ready data warehouse is an architecture for data delivery in support of the strategy of customer intimacy – the most effective way of competing in business today. It enables the management of improved overall customer satisfaction and the capability to segment customers and treat them as individuals rather than as part of a collective group. We've all heard that it costs 10 times as much to acquire a new customer as it does to keep an existing customer, but it is not good enough to stop there. It's going to cost 100 times more to reacquire a customer similar to your best customer, and it will be cheaper to let some existing customers go than to keep them. The bridge to making this work is customer segmentation supported by a CRM-ready data warehouse.

> *Incorporation of external data.* If you can conceive of a need for external data, chances are the data is available or derivable from available data. Some data warehouses are comprised of mostly external data as consortiums that clear data that continue to gain prominence in some industries. The high-value component of much of this *reverse-appended* data is its capability to generate effective cross-sell and up-sell possibilities as well as lists of prospects with characteristics similar to your best customers.

> *Realization of the architecture nature of the data warehouse.* The CRM-ready data warehouse is not a singular database. It is many databases playing roles within an integrated architecture that is multipurpose, flexible, and cross-functional in nature.

> *Unpredictable, varied, and growing access patterns.* No "one size fits all" is possible when it comes to access tools. Most best-practice programs have three to six different access tools in the hands of a varied user community. The tools span various access categories, such as relational OLAP, hybrid OLAP, multidimensional OLAP, push, desktop OLAP, reporting, data mining, data visualization, and portals.

Access patterns have risen largely due to the need for custom, not mass, approaches to the marketplace. Survey.com predicts that the average number of users of a data warehouse will be 2,718 by 2002. Much of this usage will be from suppliers, partners, customers, and employees – not just the traditional knowledge workers.

E-intelligence analytics. Web houses that take you beyond site traffic statistics and drive strategic directives of customer profitability, product profitability, customer satisfaction, and return on investment support the highest value-add of e-business to a company.

Packaged solutions. Some packaged solutions are dangerous if not supported by business process; however, incorporating some measure of packaged approaches is worth consideration for new efforts. Time to market is critical for CRM efforts. Information technology (IT) must become, practically speaking, an internal systems integrator, integrating heterogeneous system and software components from software development houses. IT must primarily be responsive to business needs. In addition to technology skills in their core area, skills required by all include customer service, requirements gathering, information collection, and planning. Understanding of and responsiveness to business needs are paramount in today's environment. Integration of best-of-breed purchased components contributes to that goal.

Analytical calculations for derived dynamic data. Customer lifetime value, promotion response, customer spend percentile, customer profitability, and customer category spending are just a few of the dynamic metrics that can be proactively and dynamically added to a data warehouse. The interesting part is the capability to track customer patterns over time.

Data quality issues abound requiring custom solutions. Since the beginning of data warehousing, data quality has been the number one risk to data warehousing efforts. This remains true today. Many overestimate the quality of data in their operational systems. Often these systems, data must be cleaned up prior to feeding the warehouse with their data. Cleanup processes often require specific company knowledge to ensure that the data is properly represented.

Financial payback for CRM data warehouse efforts. Whether going for complete ROI or stopping short by targeting intermediate factors, such as customer satisfaction, number of customers, or promotion response, it is important to have tangible goals for a CRM-ready data warehouse program.

Our challenge this year is to rise above the tsunami of shortcut and siloed approaches to data warehousing and establish a long-term program architecture comprising all of these trends. The challenge is to recognize and drive personalization opportunities everywhere possible and to ensure that the infrastructure and support services are satisfying the ever-increasing demands for our CRM-ready data warehouses.

Your data warehouse is doubling in size every year, and you suspect that you'll be adding a clickstream data source from your corporate Web site before the end of the year. Your marketing group doesn't yet know precisely how it is going to leverage the new clickstream data for personalization and customer intimacy, but it does know that it doesn't want to be the last to figure it out. Too many signs suggest it will be of strategic competitive advantage.

CLICKSTREAM VOLUME + DEMAND FOR ACCESS = BIG TROUBLE

As data warehouse manager or extract/transform/load (ETL) team leader, you know that the information that users learn from their clickstream data will stimulate additional requests for analysis. That means reprocessing and reloading the data many times before thorough data transformation and data hygiene requirements can be fixed in stone. This may not be new; after all, you've seen significant change in your ETL process for several years.

However, this is likely to be as challenging a design-as-you-go solution as you've seen because clickstream volumes may match – even exceed – current volumes, and ongoing support of existing ETL is not getting any cheaper. What to do?

The answer: start building your new ETL job streams within a scalable ETL development environment. The initial clickstream ETL job stream will change many times before you're through. There is no better time to start positioning a scalable technology that delivers in the very large data warehousing development environment.

Defining the Environment

A scalable ETL development environment addresses continuous change in your existing ETL infrastructure with programmer productivity benefits that save time and money; at the same time, you will also reduce existing support and maintenance costs. The real clincher, however, is in providing an extremely reliable and high-performance ETL processing environment that can also scale predictably to meet faster response times and higher throughput. There you have it: extreme scalability, massive throughput, robust performance, and low cost of ownership. These characterize the scalable ETL development environment.

Although many ETL tools are on the market, few deliver the scalability, throughput, performance, and low cost of ownership you will need to support an environment that integrates clickstream data with your existing data management environment. Solutions are available, however, from parallel technology vendors, such as Ab Initio and Torrent Systems, Inc. These solutions deliver three key benefits in the scalable ETL development environment for extract, transform, and load jobs:

1. Programmer productivity gains

2. Reduced maintenance costs

3. Scalable high performance

Programmer Productivity Gains

Developers always seem to take offense when programmer productivity tools are suggested for ETL. How many times have you heard someone say, "Why do I need a tool? I can write code that performs more quickly and be done with a simple transform. I've already built it in my class library, and I can code it in 15 minutes." You rarely see it turn out that way, though.

Ask your favorite C/C++ developers if they have a routine to convert packed decimal to text. Nobody would actually code one until it was needed, so you don't often find one in your class library. Ask how long it takes to translate the COBOL copybook into the input interface specification for your transform. Probably an hour – and that is if you understand the copybook format or you've already processed similar feeds.

The point is this: you shouldn't have to burn valuable development resource coding interface specifications! Rather, you should have your ETL development team tackle more elaborate ETL data and application concerns – issues that come up often and need a good strategy – such as synchronizing data among different sources, coming up with a refresh strategy, designing checkpoint and recovery logic, or even data quality data audit checkpoints.

Consider some of the tasks that would take a day or more to code and test without a scalable ETL development environment. It is no accident that most of the components to build these steps are available from tools designed for scalable ETL:

1. Formatting packed decimal/binary decimal/EBCDIC data into string text for dimension tables

2. Sorting and merging data files with logic handling for merge exceptions

3. Calculating aggregates for a single file or across many sorted and merged files

4. Selecting certain records from an ETL job stream for data quality or data auditing checkpoints

5. Extracting a 10 percent random sample from your warehouse for an ad hoc data pull

6. Removing duplicates from an ETL job stream

These are the bread and butter of ETL coding that essentially come free with your scalable ETL development environment. That is, developers could be using prebuilt and tested components to solve these problems in hours, not days.

Reduced Maintenance Costs

Let's now look at maintenance costs that require considerable resources. A scalable ETL development environment tackles the need for custom-coded C/C++ extraction and transformation logic head-on. It must alleviate the need for a large C/C++ development team, and it does.

Most of the transformation logic can be built defining transformation rules in a simple procedural language instead of C/C++. As a result, programming support costs are much less. You don't have to worry about supporting unreadable code after a developer moves on. In most cases, scalable ETL development environments provide many self-documenting features.

One of the other classic support problems with ETL performance is the number of potential interfaces between sources and targets – databases, tables, or flat files. They have a nasty habit of growing exponentially; in fact, there are potentially $n \infty (n\text{-}1)$ possible interfaces between the n sources and targets. If you have just four sources, you are already supporting up to 12 interfaces to those sources. The best way to reduce the maintenance and support costs associated with the large number of interfaces is for the scalable ETL development environment to manage these interfaces using an ETL-specific metadata database and version control.

PREDICTABLE PERFORMANCE IN AN UNPREDICTABLE ENVIRONMENT

The third and final key benefit to scalable ETL development environments derives from the performance options that are embedded within a scalable ETL development tool. It separates the developers from the details of executing their code in a parallel or distributed system.

Let's say that you have an ETL job stream in place wherein you can deliver data to the warehouse. How do you make it four times faster? Better yet, how would you process four times as much data in the same amount of time? The answer to both questions could be to buy more hardware, but let's assume that the next fastest processor is only twice as fast. You're stuck with finding a software solution.

Now compromised, the development group starts planning its code to be not only optimized running on one CPU, but its also plans to make use of multiple CPUs/disks concurrently. This in itself is not too daunting a task for more advanced programmers, and your development team might suggest a multithreaded or multiprocess programming solution in which you partition the data into four pieces on four disks and run four copies of your program on four CPUs at the same time. This is a great solution, until you realize that only 10 percent of your development team can support the code.

What if the problem changed again, and you were asked to increase the number of processing steps in the ETL job stream that slowed down the data stream by half. It's not good enough: your users are howling for the same run time! You could partition the problem into eight pieces using the same techniques as before. You would have to make sure that you added extra file systems to keep each of the eight streams fed with independent I/O, but you did it before, so you can do it again, and the server has eight CPUs.

Now your clickstream data is feeding into your existing data stream, and all is well. However, volumes are growing, and so are user requests. What do you do when you're asked to process twice as much data again, but you've run out of CPUs on your server? Do you start building programs that use message passing or remote procedure calls (RPC) to do distributed processing? This would enable you to make use of CPUs and disks on other servers. You could build scripts to move the data around to do the processing on the new server, but now only two to three percent of your development team has the expertise. What happens when they leave?

This unpleasant scenario is occurring more often than not. The key benefit from scalable ETL environments is that developers are largely shielded from the details of coordinating or even thinking about parallel and distributed programming. Using a visual programming environment, a developer can implement ETL job streams with many prebuilt components, without concern for whether a job stream will use four-way data parallelism or 16-way data parallelism. Unlike components in standard ETL tools, the prebuilt components were designed from the beginning to be run across multiple disks, multiple CPUs, or even multiple servers.

Developers are also shielded from the hardware specifics of allocating programs to servers/CPUs or moving data around for processing. The hardware allocation details are managed in a system configuration component updated as your system group adds more hardware in the form of new servers (CPUs) or more file systems (disks), requiring no application development change to make use of the new hardware.

Suddenly the growth path becomes predictable and clear for your ETL environment. As business requirements grow, you can scale the hardware to accommodate the processing requirement without requiring ETL programming or software changes because you've designed your ETL applications to be scalable from the beginning.

REAL-WORLD DEMANDS NEED REAL-TIME DATA WAREHOUSING

Real-time data warehousing is no longer a buzzword within the scalable ETL development environment. Your marketing group can extract data in near real time, perform analytics, and deliver the personalization that differentiates your business. And, you can make that happen.

On a final note, industry watcher META Group estimates that within the next year or two, data analysis efforts for e-commerce sites will need to deal with 10 times more data, analyzed five to 10 times more quickly than current best practices. That's 50 to 100 times more capacity. Are you ready for it? You can be – really.

Effective data extract, transform, and load (ETL) processes represent the number one success factor for your data warehouse project and can absorb up to 70 percent of the time spent on a typical warehousing project. ETL tools promise quick results, improved manageability, and metadata integration with other common design and implementation tools. However, due to the potentially huge amounts of money involved in a tool decision, choosing the correct ETL tool for your project can present a daunting challenge. With a bit of internal questioning in advance followed by a careful review of your key needs against the choices available on the market, you should be able to choose the most effective ETL tool for your project.

ETL tools perform, as you may guess, at least three specific functions, all of which focus around the movement of data from one place (file type, server, location, and so on) or system to another. More encompassing than a simple file copy process, this class of software generally reads data from an input source (flat file, relational table, message queue, and so on); passes the stream of information through either an engine- or code-based process to modify, enhance, or eliminate data elements based on the instructions of the job; and then writes the resultant data set back out to a flat file, relational table, and so on. As you may have guessed, these three steps are known as extraction, transformation, and loading, respectively.

You may also see these tools classified as data-cleansing tools, although here we should make a careful distinction. Although data cleansing definitively can encompass the ETL functions, most true data-cleansing tools are not designed to perform true ETL and, vice versa, most ETL tools provide only limited true data-cleansing functions.

A quick example can show the difference between the two types of tools. Suppose that you have an input data file containing the full name of a customer on an account. Particularly in financial services, this data may contain any manner of formats; but, for this example, let's use "John S. and Mary Smith." A true data-cleansing tool will possess integrated functions to parse this phrase into the two named account holders, "John S. Smith" and "Mary Smith." This type of complex string parsing is not generally a strong function of an ETL tool. The ETL tool, however, will generally be better at efficiently looking up the name "Mary Smith" in a relational database customer table and returning the integer numeric key value related to the "Mary Smith" customer name.

More than 75 tools on the market purport to have some ETL function. These tools may be classified according to function, engine type, and development environment.

Function

The tools within this classification fall into several general categories:

EtL Tools. These "small t" tools emphasize the extraction and loading processes and should actually be referred to as data-migration tools instead of full-function ETL processors.

eTL or ETl Tools. These "small e" or "small l" tools typically accept only a specific input or output type, such as a flat file source or a specific database format, but offer fairly robust transformation functions within the processing engine.

eTl Tools. These "capital T" tools perform the data transformation step relatively well but may lack efficient connectivity to many of the common data formats that may be encountered.

ETL Tools. These "complete" ETL tools provide a rich mix of functionality and connectivity but may be significantly more expensive than tools found in the other categories. For extremely complex projects or those attempting to process massive amounts of data, these tools may present the only true option for ensuring success in the ETL phase of the project. In other cases, this class of tool may offer features that simply are not required in the existing environment.

Engine Type

The engine type classification segments the tools by how the developed ETL processes are executed. Typically, the tools encountered fall into one of two categories: server engine or client engine. The server engine enables the execution of multiple concurrent jobs from more than one developer. Server engines generally take advantage of multiple CPUs and are designed to coordinate and manage the execution of multiple simultaneous routines.

The client engine is simpler and assumes that the ETL routines are executed on the same machine on which they are developed. Concurrency for multiple jobs is limited, if allowed at all, and client engine tools often do not support scaling across multiple CPU machines. Client engines are significantly cheaper to purchase due to their more limited functionality and scalability.

Development Environment

Development environments are typically split two ways: GUI-based or code-based tools. Code-based tools are the most familiar and may not be considered "tools" independent of the language they represent. For example, Perl can be used as a code-based ETL tool, but it is also a more generalized programming language. The embedded transactional code languages within common database platforms (for example, PL/SQL with Oracle, Transact*SQL with Microsoft SQL Server, and so on) may also provide ETL functionality but are not limited to this capability. Aside from general programming languages, several tools on the market use a custom-scripting language developed explicitly for the optimization of ETL routines.

GUI-based ETL tools have been on the market for at least six years. The purpose of the GUI is to remove the coding layer for the developer and enable the generation of ETL routines without requiring the mastery of any particular coding language. GUI tools are also useful in that they provide some self-documentation about the job flow just from the layout and positioning of the graphical elements.

KEY QUESTIONS TO ASK

Now that you have some background on the types of tools available, what criteria should you use to help decide which tool is best for you? We will focus the discussion around the five Cs: complexity, concurrency, continuity, cost, and conformity.

Complexity

Complexity is generally evaluated through the following series of questions:

1. How many distinct sources of information will need to be processed with this tool?

2. If using files, what is the expected size of the input files? Are they ASCII or EBCDIC formatted? For mainframe file sources, do they contain OCCURS, REDEFINES, or packed fields? For other file sources, are the files delimited or fixed length? What character is used as a delimiter?

3. If using relational database sources, how many rows will need to be queried each time the job is run? Do primary and foreign key

relationships exist within and/or between the tables of interest? Can the tool support the types of joins needed to execute against the tables of interest? Is it possible to see and tune/edit the SQL the tool is generating against the database?

4. What "magic values" exist within given data fields? Is it necessary to translate computer codes into more understandable terms?

5. Is there a need to join information together from dissimilar sources? Is there a requirement to match flat-file data with fields in a relational database table? Is there a requirement to join information from two different types of databases?

6. Does the data warehouse use the "slowly changing dimensions" concept? If so, how does the tool support appropriately updating dimensional attributes?

7. How well does the tool support testing and debugging while in development?

Concurrency

Concurrency evaluates both the number of developers and number of simultaneous processes that the tool must support:

1. How many developers will need access to the tool?

2. Is there a need to implement security to limit access for individuals or groups? How does the tool support this? Is it possible to leverage LDAP or other in-place security mechanisms for authentication purposes?

3. How are change management functions or *versioning* handled within the tool? What happens if two developers attempt to make changes to the same routine or transformation object at the same time? Can

older incarnations of various routines be retained for reference? What complexities are introduced through the specific versioning or change management functions used by the tool?

4. What is the maximum number of simultaneous processes that are expected to run? What size server would be required to support this number of concurrent processes with this tool? Can the tool support running twice that many simultaneous processes?

5. How flexible is the tool when executing dependent or sequential job streams? Can the tool recover from a failed job without manual intervention? How flexible is the error condition testing? How are errors logged?

6. Is the product multithreaded? If so, is the threading unlimited or set within the product? Are multiple jobs required to achieve high degrees of parallelization?

Continuity

Continuity involves two distinct subjects. Continuity establishes the scalability of the tool to meet future demands from system growth, and it evaluates how reliable the product is under unusual circumstances:

1. If data sources double in size, can this tool support that level of processing?

2. How easy is it to make changes to developed jobs or routines?

3. Can more than the minimum hardware and operating system requirements to support the product be deployed?

4. How well does the product respond to spurious failures, such as network problems, server crashes, or running out of disk space?

5. How often does the vendor upgrade the product? How painful are the upgrade installations?

Cost

Cost represents the real purchase cost and perhaps the hidden support costs associated with each product:

1. What is the quoted price for the product options needed to be successful with this product? Where will additional software licensing costs (adding data sources, developer seats, server CPUs, and so on) be incurred in the future?

2. What future software costs will be incurred to receive software support and upgrades from the vendor? Is the first year of support and maintenance included with product purchase?

3. Is it necessary to purchase additional hardware or software components in order to effectively use this tool?

4. Who will administer the tool? Will it be necessary to hire a dedicated person or can someone already on the team fulfill this duty?

5. How much training is required to learn the tool? How much additional training is required to administer it? What advanced topic courses are offered? What do the courses cost? How often, and where are they offered?

6. How available are support resources outside of those provided by the vendor? What resellers or systems' integrators support the product? Are there local users groups? Are technical forums available on the Internet?

7. How long should it to take to implement our first production processes? How long is the typical learning curve on a tool like this (measure to the point of effective use of the tool, not just through the end of training!)?

Conformity

Conformity outlines how the tool behaves against existing architectural limitations or requirements:

1. Does this tool support the hardware platform and operating system that will be used? Are there any incompatibilities with any other software typically run on the servers (backup software, virus scanning software, and so on)?

2. Does this tool integrate with the network and user security protocols used? What, if any, work-arounds will be required?

3. Does this tool allow access to the type and location of data sources that will be used? Should mainframe files be accessed directly (and can the tool support this?), or is it necessary to extract mainframe data to files first?

4. Can the tool import and/or export metadata or other key information with the specific tools (data modeling tools, business intelligence tools, and so on) that are deployed in the data warehouse?

5. What form of documentation does the tool generate? Does this format and content meet internal documentation standards? What additional documentation will need to be developed and maintained outside of the tool?

6. Is this purchase a departmental tool, or will the tool be used across the enterprise? How would the tool architecture scale to grow from a departmental solution into one suitable for enterprise deployment?

EVALUATING THE ETL PURCHASE DECISION

You will find that ETL tools can range in cost from free (freeware) to more than several hundred thousand dollars. As such, your decision delivers a widely ranging and potentially lasting impact on your system's architecture. If you are not already using a tool, you obviously first must compare the decision to purchase a tool against the "keep doing it like we're doing it today" option. Very few IT departments can meet the current and future development needs of their business users without substantial growth. One way to offset the need for additional head count is to take advantage of the efficiency that modern ETL tools can bring if chosen and used effectively.

To be blunt, you may encounter the feeling of buyer's remorse with almost any software purchase you make. Unfortunately, no single right answer is available in any tool choice today (despite what vendors tell you). As such, your goal should be to understand in advance where the expected weaknesses will occur with your chosen solution and be prepared to mitigate those consequences.

First Steps — Deal Breakers

Your first order of business in outlining your ETL selection is to determine whether any critical business factors exist that might help limit the number of tools you will consider. Examples of these factors may include:

1. We must have a tool that runs on {specific operating system}.

2. We must be able to process {amount} GB of data nightly in a {number} hour batch window.

3. We must have connectivity to {type of data source}.

4. We must be able to run our ETL jobs on the mainframe.

5. We must have a tool that supports concurrent development by multiple sites.

6. We absolutely cannot spend more than ${x}.

The specific objective is to identify conditions that weed out solutions which will not be acceptable within your given architecture. Ideally, this first step should identify a simple list of the top six or seven tools that might meet the majority of your requirements. An exhaustive review of the tools is not required at this point.

Making the Final Decision

After winnowing out the clearly unacceptable choices, your next step in the process is to outline what criteria you will use in the evaluation. You may use the questions in the earlier sections of this article as a guideline, but you should be fairly specific and outline specific data sources with which you require connectivity, hardware expectations, operating system options, and so on.

After building the list, generate a matrix with your decision criteria as rows and the tool choices as the columns. Then, rank each of the tools using publicly available information (Web sites, publications, and so on) as much as possible. If you do not feel that you have enough information available to adequately answer your questions, contact a vendor representative to obtain the necessary information to complete the matrix. Note that managing more than two to three vendor presentations and handling appropriate follow-up can be an extremely time-consuming task. For this reason, you may want to begin on-site formal vendor presentations after you've created your short list of the two to three finalists. This approach will allow you to truly focus your energies on understanding why one tool is specifically better than just one or two others. With a wider field, you generally will find many triangular relationships (tool A is better than tool B at criteria 1; tool B is better than tool C at criteria 2; tool C is better than tool A at criteria 3; and so on) that confuse the overall picture.

Eventually, you should arrive at a reasonable conclusion about which tool will best meet your needs. You should definitely try to obtain a customer reference from each vendor, preferably another company doing approximately the same function as your organization (although for competitive reasons this may be hard to obtain). Focus on your key technical requirements to ensure that the customer reference you obtain is relevant and useful to you. A customer reference accessing similar systems and operating on similar hardware and operating systems is probably more applicable than another similar industry reference on a completely different platform.

In the end, of course, it all comes down to putting ink to paper and taking on the responsibility for this new project. Much like raising a child, you won't know for some time how decisions you make today turn out, and you'll inevitably have second thoughts along the way (Was I ready for this?). However, with a solid decision-making process behind you, you can expect to grow and succeed with your selection. At the final bell, it is the results that count!

ROLE OF NEARLINE STORAGE

A few short years ago, data warehousing was not much more than a theory derided by most database academicians. Only a handful of brave companies experimented with the new architecture. What a difference a few years make. Today, it is conventional wisdom – a legitimate international phenomenon embraced by thousands of companies using data warehouses for mission-critical tasks in diverse industries.

With the architecture's new possibilities comes unprecedented storage growth. Prior to data warehousing, volumes of data were measured in k bytes, megabytes, and gigabytes. But with data warehouses in the picture, storage is now measured in hundreds of gigabytes and in terabytes – a word no one had even heard of before. A few very large organizations are already considering petabyte data warehouses.

Information Explosion

The growth evident with data warehousing occurs everywhere, and the growth rate is as astonishing as the growth itself. Corporations have never before faced the huge volumes of information that accompany the data warehouse.

Where is it coming from? What is fueling this fantastic growth? You can start with the appeal of data warehousing itself. Prior to data warehousing, organizations wrestled to get information out of transaction processing systems. Although some of these systems are sufficient for getting limited information into users' hands, they leave much to be desired when it comes to extracting data. In particular, transaction processing systems fail to provide two key elements:

1. *Integrated information.* Each application has its own unique understanding of data, and no two applications are the same. That means using application information to look across the corporation is not a viable option.

2. *Historical information.* Transaction processing system applications focus on very current information. However, when it comes to gathering and assimilating historical information, the applications pay little or no attention. Unfortunately, when a company evaluates factors, such as customers and customer buying habits or predicting buying trends, it is historical information that drives the process. After all, customers are creatures of habit, and history is a great indicator of future behavior.

Aside from the shortcomings of transaction processing systems, other very good reasons exist for the growth of data warehousing. One of the most exciting phenomena is the data warehouse's capability to shelter information that fosters new data uses within the corporation. Data warehousing makes it possible to take advantage of entirely new styles of applications – analytical, business intelligence, and exploration, to name a few. Using these applications, corporations are finally able to leverage their information throughout the system and to better cope with business mergers, globalization, product introductions, and business changes. Another advantage is that data warehouses store data at an atomic level, which can be endlessly reshaped to supply ever-changing informational needs throughout the corporation. The savvy organization can achieve a new level of competitiveness based on information available in the warehouse. It is no surprise that this effect on corporations planning to grow and remain strong find data warehouses at the heart of the new wave of applications.

Where Is All the Space Going?

The use of data warehousing is not all that is growing. The volumes of data inside the warehouses are growing right along with it. The reason behind this growth boils down to history and detail. Whenever the simple equation history ∞ detail arises, masses of data are an inevitable result. Also, every time an analysis is performed, a new summarization is created. Although the appearance of summarizations in a data warehouse is a sign of success, they take up space just like any other form of data. The final reason that the volumes of data within data warehousing have been growing is in preparation for future unknowns. In this world, designing a data warehouse for unknown requirements inevitably requires the designer to include data that may be used very infrequently – yet another factor contributing to the heft and size of a data warehouse.

Two Kinds of Data

Start storing information in a data warehouse, and before long, you witness a new phenomenon called *separation of data.*

As long as the total volume of data is small, there tends to be a lot of actively used data. In fact, a small database may not even have any infrequently used data. But, as the warehouse's volume of data increases, the separation of data into two classes becomes more and more apparent. At some point, the corporation ends up with more data than it can possibly actively use. Before you know it, more infrequently used data than actively used data exist – a lot more.

The two data classifications are generally caused by overestimating the importance of historical data and forgotten summarizations.

The data warehouse is the first place where users have ever had the opportunity to store and use historical data. With such an opportunity, users tend to store as much historical information as possible, not knowing when they may ever get more. The warehouse designer tries to accommodate them, and, in doing so, far more historical information enters the data warehouse than would be actively used. Summarizations are another reason for the growth of infrequently used data. Created on the assumption that people will want to look at them in the future, summaries are actually a sign of warehouse health. However, the truth is that people forget that they even exist. Even though use is infrequent, computer operations people are reluctant to migrate the data, and, as a result, the data just sits in the warehouse taking up space. These are just two of the multiple reasons for infrequently used data.

Business and Technological Challenges

The separation of frequently and infrequently used data presents business and technological challenges to any organization including high costs, sacrificing performance, and design issues.

You pay the same amount for storage, used or not. So, does it make economic sense to have a large amount of storage data and yet barely use some, and perhaps a great deal, of it? Is there any other resource in your corporation you would treat this way? Then why should computer storage – in particular, disk storage – be any different?

Companies also sacrifice performance in their data warehouse when infrequently used data is in the way of actively used information. When a query spends 90 percent of its resources sifting through infrequently used data, it is apparent that your performance could be greatly improved by migrating data in the data warehouse across other media types, such as automated tape. In so doing, you improve performance and set the stage for multiple, varied uses of that data outside the standard query process.

Design is also affected by data separation. Because designers must stretch designs to fit the limitations of disk storage, they generally have to sacrifice the detail that would otherwise add up to volumes of data.

The bottom line is that some profound implications are associated with the separation of large amounts of data. If one aspect of data warehouses demands to be managed, this two-class separation ranks at the top of that list.

A BIRTH OF NEW DATA USES

When data separates, something very interesting happens: New knowledge management users as well as new uses appear. This secondary effect of infrequently used data can have an enormous impact on a corporation, giving way to new applications, such as data mining, predictive modeling, trending, and simulations. In short, infrequently used data opens the door for the corporation to examine aspects of information it has never been able to analyze before. Not only does infrequently used data end up being used, it is used in new ways that go beyond the context of the requirements driving the data warehouse in the first place.

Becoming aware of the issues of data warehousing and managing large volumes of data inevitably leads to the question, "How much infrequently used data do I have in my data warehouse?"

The fraction you calculate for the amount of active versus infrequently used data is a great indicator of the number of resources wasted by storing infrequently used data on disk. For really large data warehouses, it is not unheard of for this fraction to exceed 99 percent. But, any time the fraction exceeds even 80 percent, you can realize tremendous savings by looking at a hierarchical approach to storage management.

Hierarchical Approach

Part of the beauty of the hierarchical approach to storage management is the amount of money you can save. For every dollar of disk storage, you could have the same amount of storage for 42 cents by choosing photo optical storage, or seven cents if you choose siloed tape storage. In other words, you can realize more than an order of magnitude of savings by planning a data warehouse that places data on both disk and nearline storage.

Take a very simple example. Suppose that you have a two-terabyte data warehouse on disk storage for which you paid $25 million. (Hypothetically, this covers the full cost of the warehouse, including transformation, middleware, storage, and processor costs.) If you put the same data and the same warehouse on a combination of disk and nearline storage, you would have paid something in the order of $3 million. That is some savings.

The hierarchical approach also improves performance. When you migrate infrequently used data to nearline storage and save money, performance gets better, not worse. When any query is run against a large, all-disk data warehouse where the data has separated, the query has to plow through tons of unneeded data to find the data you actually need. Whether this means sifting through indexed data, using full or partial table scans, or using other data access means, the fact is that when you have a lot of infrequently used data, actively used data gets buried. As a result, you can get a tremendous performance boost just by moving infrequently used data to nearline storage.

As long as the data warehouse is limited by the constraints of a purely disk storage medium, it will never fulfill its true potential. However, with the combination of disk storage for actively used data and nearline storage for infrequently used data, it is economically and technologically possible to extend the storage boundaries.

Cross media management software can transparently extend to the relational database management system's (RDBMS) SQL access to the complete hierarchy of storage. This approach removes the technological handcuffs and costs constraints of a disk-centric approach to data warehousing, dramatically increasing the business value of your data warehouse.

A simple formula for determining the amount of infrequently used data in your warehouse can be found in the original Inmon white paper, located online at http://www.storagetek.com/Datawarehouse, courtesy of Storage Tek's Data Warehousing Division.

Building a data warehouse (DW) in today's new global economy can be very challenging. Although the conceptual DW design and architecture of a traditional brick-and-mortar company versus a dot-com may be very similar, it is the complex and dynamic nature of the dot-com organizational environment that can be the greatest challenge.

Explosive Growth

Internet-based companies, by definition, typically function in an environment of volatility, both within their operations and with the technology they use to keep up with business growth and market demand. They operate at an accelerated pace in response to the ever-changing business environment, competition, and customer demands. To gain competitive advantage, dot-coms must be willing to take risks, implement bleeding-edge technology, change direction quickly, and push existing technology to the limits – all important considerations and challenges for DWs.

Also, executives at these organizations recognize the importance of information; they understand that in a time of fierce competition, the company that is the most effective at leveraging information will pull ahead of its competitors. As a result, these firms are beginning to plan for a DW early in the start-up and launch stages of the company when growth, complexity, change, and uncertainty are most prevalent.

Unfortunately, when a dot-com gathers the DW's requirements during its start-up phase, it may not have the luxury of any source systems or data to drive those requirements. The requirements gathered for the DW may then provide the requirements for source systems, a bit of a role reversal from the traditional approach.

During the lifecycle of a DW implementation, project deliverables must be flexible and extensible, changing virtually on a weekly basis as the business grows and operational systems are enhanced or upgraded with the latest technologies. This rapidly changing environment, as well as the dot-com's need to respond quickly to new business opportunities, requires that the DW implementation team be extremely resourceful and adaptable. The explosive organizational growth within a short period of time, coupled with a rapidly changing operational environment, makes DW planning and sizing complex.

Another important challenge that we are seeing associated with Internet start-ups is the need for speed. The demand for rapid implementation and quick results contributes to extremely aggressive schedules and short implementation windows that do not allow for mistakes or miscalculations. The DW has become an important enabler for dot-coms. As such, the DW must be in place and must be flexible and scalable enough to keep up with the pace of the business. Because of this, Internet companies are very interested in prebuilt analytical applications, supported by a data warehousing model and in technologies that enable them to get their DWs up and running in an accelerated timeframe.

At the same time, these firms understand the importance and logic of supporting an iterative approach to data warehousing. This is the best way to balance and achieve their requirements relative to timing, flexibility, extensibility, scalability, new technologies, and so on. As such, dot-coms typically prefer multiple product releases in short intervals versus the single release (big-bang) approach, even if all requirements are known up front, which is rarely the case. Although the implementation window is extremely small, the speed with which the business changes, as well as the dot-com's willingness to use new technology, forces change in all phases of the DW implementation project. A sound change control process and a highly integrated team with the ability to provide high-quality work and deliverables on accelerated timetables are, therefore, necessary for a successful implementation.

The Dot-Com Organization

Dot-coms typically contain a group of IT zealots, who are technically advanced, extremely demanding, and in short supply. Working at a brisk pace to develop tools, warehouses, systems, and processes that support their organization's business and competitive needs, these professionals must be both capable and resourceful. It is not uncommon for professionals in this type of position to work extremely long hours – sometimes through the night – to enter new markets and expand the business *footprint.*

This type of working environment places pressure on the DW implementation team to meet the ever-changing business and user requirements and to remain informed about the strategic and tactical business implications to the DW design.

On a positive note, many executives are technically savvy individuals. They are becoming increasingly aware of recent technology advancements; the value of information; and how, when integrated through a DW, the flow of useful analytics related to customers, products, competitors, vendors, and markets for competitive advantage can be increased. These executives are willing to assume risks to implement a new technology solution, and it is not unusual for them to expect immediate results – another challenge for the DW implementation team.

Technical Architecture and Implementation

In the high-growth and dynamic dot-com environment, the architecture of the DW must be both highly scalable and flexible to accommodate new subject areas and operational systems with minimal impact to the existing data warehouse environment. It must also be highly sophisticated in order to address significantly higher data volume associated with clickstream data, which is different from the traditional data transformation associated with legacy systems. The extract/transform/load (ETL) component of a dot-com requires building a DW architecture that addresses the operational problems of many diverse source systems, including new sources of information, such as the Web, syndicated data, and wireless technologies.

SUMMARY

The next generation of data warehousing will be driven by a few important themes: time to market; the tremendous increase in data volume brought about by the need to store granular customer information, including clickstream data and e-analytics; and the need to make this information available to all customer touch points. Therefore, DW practitioners and product vendors must leverage their experience, methods, industry expertise, and technology to develop prebuilt, integrated knowledge-based solutions and analytical applications to be successful within the virtual walls of today's dot-coms. Furthermore, to create a DW that acts as the enabling foundation and leverages information effectively throughout the organization in order to create sustained competitive advantage, the implementation team must combine both business and technical skills. At the same time, the team must be creative, flexible, and – most of all – adaptable to working in relatively unstructured environments where change is the norm.

Although new-economy companies certainly present some unique challenges for these individuals, the satisfaction that can be derived from implementing data warehouses that deliver real value to start-up enterprises is unparalleled.

Nearly all organizations planning to implement a customer relationship management program will need a data warehouse or decision-support system (DSS).

But companies should not expect that a data warehouse and decision support system will immediately permit them to forecast the future or assist in developing a corporate strategy. Using a DW/DSS for customer relationship management (CRM) is a learning process. The learning developed in one stage forms the basis for more advanced analysis and applications. See Figure 8.1 for an illustration of the evolution of CRM using DSS/DW.

This discussion of the evolution of DW/DSS use concentrates on the characteristics of use at each developmental stage and the questions asked by people using the DW at different stages of the process. The stages are also helpful in positioning technology, applications, data, and the integration or use of information resources.

CHAPTER 9

Types of DW/DSS

Topics in this chapter:

- DW/DSS Type One: Reporting

- DSS/DW Type Two: Analyzing

- DSS/DW Type Three: Predicting the Future

- Types of DW/DSS – A Summary

- Stumbling upon Hidden Business Rules

- Discussing ERD v/s Star Schema

DW/DSS TYPE ONE: REPORTING

The first type of DSS is characterized by a large quantity of *predefined queries.* Technologists set up queries after receiving requests from business users, or the management determines the queries to communicate what needs to be known after a period or a process is completed. These types of reporting systems usually provide a complete set of charts/graphs/cubes about a specific area of the business. These graphics answer the most frequently asked

A - Predefined
B - Ad Hoc Queries
C - Analytical Modeling
D - Event Driven

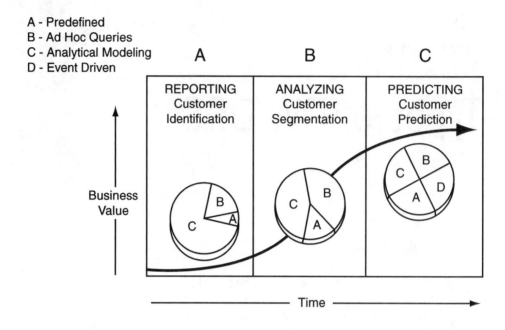

FIGURE 9.1.

Evolution of CRM using DSS/DW.

questions about the company, its market, or customers. Section A of Figure 9.1 shows the distribution of use types and the associated business value.

Reporting is the major function of a DSS in the early stages. The underlying queries are known, and the data is mostly summarized and presented quickly. This stage of a DSS answers the strategic question, "What happened?" (What may be called *hindsight viewing.*) Some typical beginning data warehouse queries are:

What are the total revenues, sales, expenses, volumes, or products produced?

Where did most of the sales, revenues, deliveries, or services occur?

What are the comparisons to or differences from past period(s)?

What are our most and least productive resources (money, products, transportation, people)?

The initiation of new levels of information and reporting has its benefits, because this information

- Provides easier access to previously inaccessible data

- Focuses on elements of information that are known by the management requestors but may not have been previously delivered or accessible

- Generates initial awareness of actions and problems

- Defines extensions to standard reporting systems

- Elevates the needs for more data and the transformation of data into information

- Opens the eyes of several key managers on what they possibly could learn if they invested more in the data warehouse

Some management attitudes may inhibit future growth and maturity in the initial stage or early uses of a DSS and DW. One such attitude is over-expectation and the desire to see all of the data on all customers in one system at one time. Users must realize that the initial implementation is only a fraction of the actual potential volume, quality, credibility, creative usage, and return on investment possible from the DW.

Additionally, some people may think that the wonderful new graphical or creative display of information is the data warehouse itself or is the truthful exposition of a customer relationship. This is rarely the reality, and the management team must understand what is being achieved and what must be achieved in the future.

The initial focus should be on building the framework, the *infrastructure* for getting initial results and reports, allowing for expansion of information levels and applications in future stages. Just learning real transformation and database normalization techniques is difficult enough for the information technology and management teams.

Solving such business and data issues as "What is a customer?", "What is a product?", or "What is a channel to our customers?" can keep an experienced management team busy for months. Adjudicating the characteristics among the various business divisions, applications, and databases can shorten this process.

Figure 9.2 shows the distribution of the reporting types of DW/DSS usage, which include some ad hoc queries and possibly some analysis applications.

When you use the DW for reporting in customer relationship management, the focus is on defining the characteristics and habits of your customers. Some of the initial questions are

Who are our customers? (age, income, gender, group)

Where do they live? (geography, economics, styles, etc.)

What did they buy in the past? (historical views)

How did they buy it? (financial transaction information)

Which ones are the most profitable? (known margins, etc.)

What is the cost to support them via their chosen channel?

Which groups of customers buy similar products?

What is the average customer revenue? Our expenses?

What is the annualized customer churn rate?

Reporting applications provide some answers found in many businesses' databases, but data warehousing provides new views and an ability to use combined, cross-organizational, detailed data to understand the past.

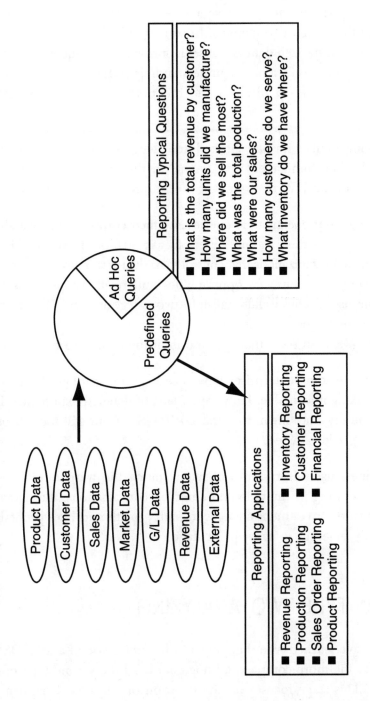

FIGURE 9.2.

Reporting applications and questions

Stage one represents the core essentials – and the most immediate data needs – of the users' requirements. In some implementations, this stage is restricted to summarized data, because the providers or requestors (business management) have only a partial understanding of the real value of detailed data. They have yet to discover the full potential of the data in their repository.

Some organizations with only limited knowledge of the actual potential of data warehousing limit their investments in the early stage to prove the initial value of the DW or to deliver on the promise of better data.

Some companies predefine their queries into their corporate data warehouses *and* do not permit their business users to ask ad hoc queries of historical customer data. Organizations that control their data resources limit the potential of creating new business opportunities and relationships. Reasons for limiting the use of data usually fall into two categories: the IT department wants to maintain good security, quality, performance, or costs; or a business department wants to own, manage, and limit the data's use by others.

Creative and leading enterprises have a strategy for getting information to users or making it available. However, a lack of detailed experiences in the user community may mean that the DW/DSS system still focuses on the familiar or predetermined.

After business users learn ways of interrogating the warehouse's detailed data and accessing its multiple and complex combinations, the magnitude of the return on investment changes to high growth and high profitability. This usually emerges in subsequent stages.

DSS/DW TYPE TWO: ANALYZING

Once we have learned what happened in the initial use of a DW/DSS, we move on to the more complex, ad hoc queries of the second type of DSS: analyzing. This stage focuses on the question of why did it happen. (See Figure 9.1, Section B.)

This organizational process enables us to understand the factors that brought about the results discovered earlier.

This transformation in understanding the value of the data warehouse is important. As shown in Figure 9.3, the use of customer information now accelerates the ability to segment and analyze customers and their actions. In addition, the types of questions become much more sophisticated, and the capabilities of the information environment are better understood throughout the organization. "You mean I can ask any question of that new DW system?" is a typical comment. Some typical data warehouse ad hoc business queries that arise in this phase are

Why did our team not meet or exceed its forecast or goals?

Why were volumes so low or deliveries later than expected?

What caused the most positive results or highest margins?

Where do we actually achieve our best ROI?

Why are inventories or resources not moving well?

This type of DW/DSS encompasses data mining through models and detailed mathematical correlation. It also can drill down into a database for minute details and come away with deductive conclusions based on the data. Business users discover trends and patterns not readily apparent from the straightforward reporting of the earlier stages. Awareness of system capabilities and enabling people to ask a wider range of questions drive a new desire to use the information infrastructure and begin to change people's thinking about the system's purpose.

Stages three and four, CRM analysis, focus on understanding customers:

Why is average customer revenue down?

Why is annualized customer churn so high?

Why did the campaign not meet plan?

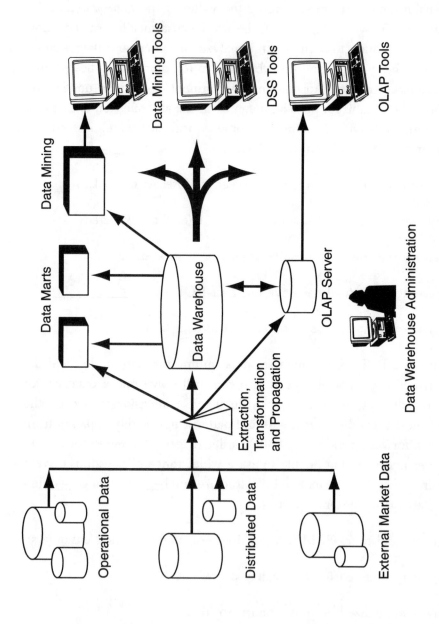

FIGURE 9.3.

Analyzing applications and questions

Why are sales of a product below plan?

Why did they buy it from you?

The reporting applications now need to interact with one another to reveal a new reality. These types of applications use more sophisticated tools, but the hallmark of the change is the use of analysis methods and models to answer the questions surrounding, "Why did it happen?" This move beyond reporting is significant, because detailed, historical data is now exploited to understand much more about past behaviors and other characteristics formerly unknown to management.

This ability to understand the past is the key to understanding the future, which is the hallmark of stage three.

DSS/DW TYPE THREE: PREDICTING THE FUTURE

Forming knowledgeable, high-percentage predictions is a specialized skill that truly separates leading companies from the rest of the pack. Those who can anticipate and capitalize on trends before they become common knowledge have an obvious edge in the marketplace.

A comprehensive data warehouse, having an analytical modeling capability, in which the queries ask, "What will happen?", provides immense abilities to achieve a prophetic facility (see Figure 9.1, Section C).

The more mature stages provide the pathway to the highest profitability and high ROI. Some of the questions that stage three (with its corresponding predictive applications) can help answer are

Which customers are at risk of leaving? (Customer retention application)

What products or services will the customer buy? (market segmentation)

What is the best way to reach a customer? (channel optimization)

How will a new product sell? (demand forecasting)

The applications are now very sophisticated and use advanced techniques of decision support, parallel query functions, massive detailed historical data, cross-functional information pertaining to customers, finite knowledge of behaviors, scoring of credit, payment ability, behaviors, propensities, pure prediction, and complex strategic decisions.

After a modicum of maturity in information management has been achieved, the enterprise becomes more mature, networked, positioned, integrated, knowledge-based, and highly flexible.

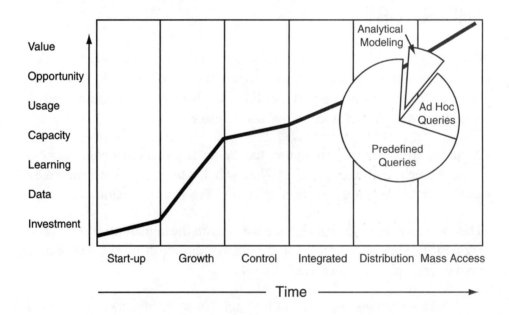

FIGURE 9.4.

Stages of development and prediction capabilities

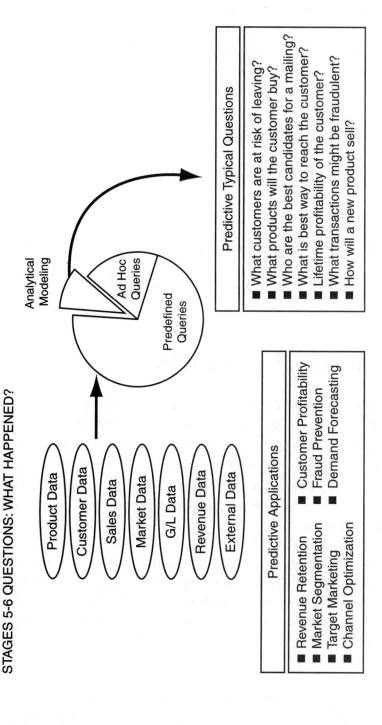

FIGURE 9.5.

Predictive applications and prediction questions

The enterprise can now get answers to questions such as, "What will happen in the future?" This is the type of DSS or data warehouse that is characterized by the applications and questions in Figures 9.4 and 9.5.

Type three of DSS with CRM – thinking like your customer – includes questions such as

Which customers are more likely to leave?

Which customers are more likely to buy?

Are they likely to defect?

What's the impact on profit when I change price?

What's the best channel to reach a specific customer?

TYPES OF DW/DSS — A SUMMARY

The most powerful use of the data warehouse and CRM applications is in the mode of active data warehousing in which the institution of real-time, event-driven, predictive capabilities are focused on action-oriented communication to customers from the marketing or services organizations of a firm. This advanced stage of a DW brings forth a highly profitable result and forges the use of relationship technologies in ways not even perceived in the first five stages.

Predefined queries and reporting are certainly the foundation of a corporate data warehousing and CRM system. However, the analytical and predictive applications are the bellwethers of competitiveness. These kinds of queries soon become indispensable elements in any company's ability to use and learn from information.

A successful enterprise uses the data warehouse and CRM to achieve a triple view: knowing the past, analyzing the present, and predicting the future (with a high rate of accuracy).

A few years ago, I was an executive at a large health care company. Every month, the officers and directors would meet to discuss matters of strategic importance: new product concepts, innovative marketing concepts, achieving customer intimacy. One month, in the midst of our strategy session, our nontechnical president blurted out an interruption to these lofty discussions that deflated our balloon of global dominance – "Before we launch another product strategy, just tell me how many members do we have in our health plan?" This simple question yielded an answer from each department, but each department answered with a different number.

STUMBLING UPON HIDDEN BUSINESS RULES

The president's focus then turned from curiosity to frustration. "How can we run a billion-dollar company effectively when we can't even agree on something as simple as the number of members (or customers) we have? Why does each department executive give a different answer?" The answer is that each executive calculated the number of members based on a hidden business rule that they didn't explain with the answer. Marketing counts a member as soon as a customer proposal is approved; sales counts a member after the first premium payment; and underwriting counts a member as soon as the health plan is at risk to provide payment for a claim. Each scenario has a different business rule to arrive at the correct answer.

This example points out a key difference between OLTP systems and OLAP used in a data warehouse – a difference that is important to building a DW solution. In OLTP systems, administrative business functions and rules are predictable and well defined. In a DW used for OLAP, many of the information requests are not known in advance and, as such, are not well defined. In fact, data mining in the DW may involve business rules and calculations that can be discovered only "on the fly."

Business Rule Discovery

A DW usually will discover business rules during two processes: data transformation and data access. Data transformation occurs first – mapping the source systems to the target DW tables. A word to the wise: Never assume that source data has integrity. We once built a DW that calculated the average length of stay (ALOS) in a hospital chain, something that was difficult to do without a DW because the data had to be combined from many disparate systems. The initial calculation produced an ALOS of negative five days. The DW was a piece of junk, or a *data outhouse*, until it was discovered that the source system was passing bad data that caused the problem. How?

The DW calculated the ALOS based on a business rule ALOS = SUM ALL DISCHARGE DATE – ADMIT DATE. This worked fine, except in some cases, when the DISCHARGE DATE was not entered in the source system, even though the patient was discharged. The source system defaulted the DISCHARGE DATE = BLANKS. Subtracting any ADMIT DATE from a blank DISCHARGE DATE yielded a large negative number, which skewed the total results. We added a business rule to the data transformation layer to check the integrity of the source system values prior to loading the DW. As Ronald Reagan said during the nuclear disarmament talks, "Trust, but verify."

Data access business rules discovery should begin with a definition of the basics of the business. How does everyone define total sales? That definition seems straightforward: The dollar amount of products shipped to customers, right? Well, usually, but should we exclude returned items? Okay, then TOTAL SALES = PRODUCTS SHIPPED – RETURNS. What about checks that bounce? Then TOTAL SALES = PRODUCTS SHIPPED – RETURNS – BAD DEBTS. And, what if someone gives away some products as a promotion? Then TOTAL SALES = PRODUCTS SHIPPED – RETURNS – BAD DEBTS – PROMOTIONS. When describing total sales, the explicit business rules should be stated.

Better Living through Business Rules

One of the primary goals in building a DW is to provide easy access to integrated information. However, if two people ask a similar question and get different answers, then no one will believe or use the DW. The discovery, encapsulation, and sharing of business rules make the use of the DW more predictable, accurate, and efficient for providing information to the enterprise. And, most importantly, when two users ask a question of the DW, they will get the same answer – based on the business rule that is used to resolve the question.

DISCUSSING ERD V/S STAR-SCHEMA

We are constantly told that things are changing ever more rapidly. Before the latest book is dog-eared, authors of newer wisdom are wanting to declare old books road kill. The PL/1 that I studied in school is just about worthless, but I don't see anybody writing that the "do-loop" is dead.

Of course, technology is only about five percent of our job. Ninety-five percent is dealing with each other. One thing I know has not changed is that people believe that which we want to believe. You may work with those who want to believe that they will someday be able to just wave their hand and create application masterpieces. They read the ads and skim the articles. I know they won't stop chasing the mechanical rabbit long enough to appreciate the road map to successful data warehouse development.

The question rages: Now that we are doing data warehousing, can we dump our entity relationship diagramming (ERD) projects, tools, and diagrams and just develop star-schema diagrams?

We hear this question more and more from teams developing data warehouses or data marts.

Let me give you the answer first: The ERD and the star-schema diagram do not compete with each other within the scope of a data warehouse project. There is a real and valuable purpose to be achieved by the use of each of these techniques. The challenge is to know how to develop each and what value to expect from each.

We still need to use entity relationship modeling – data modeling, which is analysis – to document the business information requirements for data warehouse projects. Inmon calls these business information requirements the atomic level data for the data warehouse. The definition of this atomic level data is critical to the success of any data warehouse.

The well-known star-schema proponent, Ralph Kimball, admits that "entity relationship (ER) modeling is a powerful technique for designing transaction processing systems in relational environments." The value of the entity relationship model is achieved in modeling enterprise data. Gaining consensus within the business as to the name, definition, and business rules about the data within the enterprise is the first step for anything you choose to do with that data, whether that data is used by a transaction processing system or a data warehouse system.

The data warehouse development process includes many terms for the databases to be developed during the project, including information warehouse, info mart, operational data store, data warehouse, and data mart. For the scope of this discussion, I will use *data warehouse* as the generic term and not get mired down in the subtle differences between each of these variations.

Proponents of the star-schema-centric data warehouse development recommend specifying fact tables and then specifying dimension tables. These steps are important and should be completed before you develop the physical database for a data warehouse. However, you must first have agreement on names, definitions, and business rules of the data and understand the business rules used for aggregations and calculations before you can populate the star-schema. The star-schema structure in itself does not provide the necessary information about the atomic level data required for the fact tables within the star-schema. This necessary information is the very foundation of the data warehouse. This business information is gathered through cross-functional information gathering sessions and can be recorded with the use of the ER diagram, the visual representation of your data modeling.

We work with our clients to identify a team of cross-functional representatives who know their business and their business information requirements. We develop entity relationship models with this group. We do not impose the rigorous rules of ER modeling on these business people. We start by drawing a series of boxes and asking what *categories* of information we need to know in order to be effective with the business. We name the boxes (*entities*) based upon their input. If necessary, we will work with the team to develop correct entity names (singular noun or adjective noun form). We then develop definitions for each category (entity) so that all participants can agree with the business definitions. During this process, we often identify additional entities. For example, the sales force may talk about selling "products;" however, the manufacturing team may talk about making "parts." We must determine whether these are synonyms. Should we record data about product, part, or both?

Naming and defining the pieces of data recorded for each data category (that is, attributes) begins after we have a good start on identification and definition of entities. We do this by asking a series of questions, such as, "What information do you record about your customer?" We follow a similar technique in making sure that all cross-functional team members can agree upon the documented business data names and definitions. A similar approach is used to document the business data relationships on our ER diagram. We do differentiate between an ER diagram and an ER model. An ER diagram is the graphic notation of entities, attributes, and relationships. An ER model is the ER diagram plus the associated standards-compliant names, enterprise acceptable definitions, and appropriate business rules for the business data.

There is more to developing an entity relationship model than just drawing the diagram. If you have not had the opportunity to obtain education in ER modeling, we suggest you include a methodologist as a member of your data warehouse team. The role of this person will be to take the initial input from the cross-functional team and document it in a rigorously correct ER model to properly convey the requirements to the technical team who will build the data warehouse database. (Actually, we suggest you get as many members of your organization as possible to learn data modeling; it is important to know how to think about data clearly.)

Some star-schema proponents criticize the use of an ER model when working with business people. We do not try to work with an entire enterprise ER model when working with the cross-functional team. We work with individual views of data, such as the information needed to identify and record a new customer or the information required to record a new order. Some people may refer to this partial view as a data subject area, but to the cross-functional business team, it's just a part of their business. By building and validating the ER model in the bite-by-bite approach, the business people can focus on the business issues and not be bored to tears by the complexity of the ER notation. Data modeling is not about drawing pictures; it is about analyzing how the business understands and uses data.

When you have documented the (atomic) business information requirements that fulfill the scope of a designated project – when you have finished your data modeling – you are in a good position to begin documenting the requirements for your data warehouse. The primary critical success factor for a data warehouse is that it must be built to satisfy business requirements, using data defined by and understood by the business people who will use the information obtained from the data warehouse. If you are building a data warehouse only for technology reasons, the business partners may never understand its content or value. Only by making data modeling the foundation of your data warehouse (or any information system, for that matter) will you be able to build a business tool rather than a technology nightmare.

Traditional high-level tasks to be performed on a data warehouse project include the following:

- Define the business data requirements.

- Define the system of record source.

- Define the data warehouse tool requirements and the vendors.

- Map system of record data to data warehouse database.

- Program the queries to support reporting requirements.

- Extract, scrub, aggregate, and load the data warehouse.

- Initiate the load of historic transaction processing systems data.

- Initiate the process to support ongoing load and use of the data warehouse data.

(This is a generic task list; you should customize the list as necessary.)

We propose that you can develop a project plan in which you effectively accomplish the objective of each data warehouse development task through the use of the correct diagramming technique and tools. See Table 9.1.

Ralph Kimball has developed a systematic set of nine steps to accomplish the fourth task, designing the data warehouse database:

1. Choose the process.

2. Choose the grain.

3. Identify and conform the dimensions.

4. Choose the facts.

5. Store recalculation in the fact table.

6. Round out the dimension tables.

7. Choose the duration of the database.

8. Track slowly changing dimensions.

9. Decide the query priorities and query modes.

TABLE 9.1.

Recommended Diagramming Techniques

Task Number	Task Name	Recommended Diagramming Technique
1.	Define business data requirements.	Entity relationship diagram
2.	Define system of record sources.	A matrix of source systems to target systems
3.	Define the DW tool requirements and the vendors.	Identify DBMS, MDDBMS, data conversion and aggregation as well as reporting (OLAP) tools
4.	Design DW database.	Star schema diagramming tools
5.	Map system of record data to DW database.	Any of the "industrial strength" extract,transform, and transport (ETT) tools
6.	Extract, scrub, aggregate, and load the data warehouse.	Any of the 'industrial strength" extract,transform and transport (ETT) tools
7.	Program the queries to support reporting requirements.	A good OLAP engine

An entity relationship model will assist you in the development of a quality star-schema diagram and provide value in the completion of at least task numbers four, five, and six by ensuring the use of well-documented names, definitions, and business rules for the data you put into the data warehouse and provide to your business.

The entity relationship model and the star-schema diagram are both valuable, and when used together, will enhance your ability to develop data warehouses that meet information requirements of the business. These two techniques should not be viewed as competing methods; they are companions that each serve a role in the development of data warehouses to support your organization.

What Is ODS?

Topics in this chapter:

An operational data store (ODS) is one part of the data warehouse.

Although data warehouses are widely adopted, most fail to tap the business intelligence offered by text. To date, the focus has been on developing data warehouses geared to support primarily numeric data, and the payoff has been enormous. Enterprises now have at their disposal a suite of proven practices and methodologies, along with mature tools for number-centric data warehousing. It is now time to focus on the value of text to a business and the role of text-mining techniques in harnessing this relatively untapped source of business intelligence.

WHY BOTHER WITH TEXT?

Two primary reasons to take on the challenges of text for business intelligence exist. First, far too much critical information remains inaccessible in documents. Business intelligence systems driven by data warehouses excel at telling us what happened when, but they are not very good at answering why. We easily can discover that a product's sales margins decreased by 15 percent in the last quarter in the southeast region without knowing the cause. Did a competitor release a higher-quality, lower-priced alternative? Were the margins sacrificed on this product as part of a cross-selling campaign? Did the manufacturer license another distributor in the Southeast, thus creating competition?

The answers to these and other questions are buried in documents ranging from e-mails, status memos, news stories, and press releases to complex documents such as marketing campaigns, contracts, regulatory agency filings, and government reports. To extend the depth of business intelligence, text must be considered.

Second, traditional document and text management tools are inadequate to meet the demands of business intelligence. File systems provide crude searching and pattern matching utilities. Document management systems work well with homogeneous collections of documents but not with the heterogeneous mix that knowledge workers face every day. Even the best Internet search tools suffer from poor precision and recall. (*Precision* is a measure of how many documents returned from a search actually meet the intended query criterion. *Recall* measures the percentage of documents returned versus how many should have been returned.) Finally, documents are spread across platforms in different formats and languages with little useful metadata about the content of the documents. This same type of dispersion of data is a driving factor in the development of many data warehouses. Business intelligence users need, and have become accustomed to, an integrated view of their organizations without regard to the original source or distribution of the raw data. Logically, text is just another medium for conveying information and, thus, belongs within the realm of business intelligence systems.

However, text is different. It is not structured like the numeric measures with which we are accustomed to dealing, or is it? Although text is often described as unstructured, that is far from the truth. Language is richly structured at multiple levels as linguists have aptly discovered. Structural principles are found in the formation of words (morphology), the creation of grammatical sentences (syntax), and the representation of meaning (semantics). Even higher levels of structure can be found in discourses and conversations as described by speech act theory. If we can analyze the structure of a language, we can extract the information conveyed by text. Fortunately, after decades of foundational work in computational linguistics, tools are now available to delve into the complex structures of text and to extract vital business information.

TEXT MINING: THE BASICS

Text mining is the study and practice of extracting information from text using the principles of computational linguistics. Certainly, AWK, grep, and other pattern-matching tools can extract information from text files, but these do not fall within the realm of text-mining tools. For our purposes, the key areas of text mining include

- Feature extraction

- Thematic indexing

- Clustering

- Summarization

These four techniques are essential because they solve two key problems with using text in business intelligence: They make textual information accessible, and they reduce the volume of text that must be read by end users before information is found.

Feature Extraction

Feature extraction deals with finding particular pieces of information within a text. The target information can be of a general form, such as type descriptions or business relationships. Identifying Alpha Industries as a corporation is an example of the former, although Alpha Industries, a wholly owned subsidiary of Beta Enterprises, and Margaret Johnson, president and CEO of Gamma Group, Inc., are examples of business relationships. Feature extraction can also be pattern-driven. For example, applications analyzing merger and acquisition stories may extract names of the companies involved, cost, funding mechanisms, and whether or not regulatory approval is required.

Thematic Indexing

Thematic indexing uses knowledge about the meaning of words to identify broad topics covered in a document. For example, documents about aspirin and ibuprofen might be both classified under pain relievers or analgesics. Thematic indexing such as this is often implemented using multidimensional taxonomies. A taxonomy, in the text-mining sense, is a hierarchical knowledge representation scheme. This construct, sometimes called ontology to distinguish it from navigational taxonomies, such as Yahoo!'s, provides the means to search for documents about a topic instead of documents with particular keywords. For example, an analyst researching mobile communications should be able to search for documents about wireless protocols without having to know key phrases, such as wireless application protocol (WAP).

Clustering

Clustering is another text-mining technique with applications in business intelligence. Clustering puts similar documents into groups according to dominant features. In text mining and information retrieval, a weighted feature vector is frequently used to describe a document. These feature vectors contain a list of the main themes or keywords along with a numeric weight, indicating the relative importance of the theme or term to the document as a whole. Unlike data-mining applications, which use a fixed set of features for all analyzed items (for example, age, income, gender, and so on), documents are described with a small number of terms or themes chosen from potentially thousands of possible dimensions. For example, a news story about Malaysia trade policies might feature a vector as illustrated in Table 10.1. Matrix 2 provides an example of a feature vector for an article about the Euro.

Although the two vectors share a dimension in common, most are different. The result is that unlike the relatively dense dimensional models in OLAP applications, dimensional models for documents are extremely sparse.

There is no single, best way to deal with document clustering, but three approaches are often used: hierarchical clusters, binary clusters, and self-organizing maps. Hierarchical clusters use a set-based approach. The root of the hierarchy is the set of all documents in a collection, and the leaf nodes are sets with individual documents. Intervening layers in the leaf nodes have progressively larger sets of documents, grouped by similarity. Binary

Table 10.1.
Feature Vector for Story on Malaysian Trade Policies

Intl. Trade	Tariffs	Southeast Asia	Currency Exchange	Shipping	Industrial Manufacturing
0.94	0.91	0.89	0.84	0.76	0.64

Table 10.2.

Feature Vector for Article on Euro

Currency Exchange	European Union	U.S. Dollar	Japanese Yen	Trade Deficit	Equity Markets
0.96	0.93	0.54	0.52	0.48	0.23

clusters are similar to k-NN clusters in data mining. Each document is in one and only one cluster, and clusters are created to maximize the similarity measures between documents in a cluster and minimize the similarity measure between documents in different clusters. Self-organizing maps (SOMs) use neural networks to map documents from sparse high-dimensional spaces into two-dimensional maps. Similar documents tend to the same position in the two-dimensional grid.

Summarization

The last text-mining technique is summarization. The purpose of summarization is to describe the content of a document while reducing the amount of text a user must read. The main ideas of most documents can be described with as little as 20 percent of the original text. Little is lost by summarizing. Like clustering, no single summarization algorithm exists. Most use the morphological analysis of words to identify the most frequently used terms while eliminating words that carry little meaning, such as the articles *the*, *an*, and *a*. Some algorithms weight terms used in opening or closing sentences more heavily than other terms, and some approaches look for key phrases that identify important sentences, such as *in conclusion* and *most importantly*.

With these techniques in hand, it is time to turn to the issue of integrating text in the data warehouse.

EXTENDING THE WAREHOUSE

Extending the data warehouse to support documents and text mining will require new data structures as well as new tools.

Accommodating text in the warehouse requires support for the text itself along with its metadata. Storing documents is not a problem for RDBMSs that support binary large objects. Some, such as Oracle8i, provide direct support for documents in the warehouse.

Documents are metadata-intensive objects. In general, the data warehouse should support metadata about document source, analysis, and content. Source metadata describes where a document originated and when it was loaded along with quality and timeliness information. Analysis metadata drives the type of text mining performed on documents. For example, e-mails should not be summarized, but they are good candidates for clustering using self-organizing maps. Content metadata should include at least the attributes delineated in the Dublin Core, a metadata standard for Internet resources. The Dublin Core includes title, creator, subject, description, dates of publication, copyrights, format, and relationships to other works. Content metadata also will include information mined during text analysis, such as features and business relationships mentioned in the text.

Working with text requires additional tools. Although some features are built into database systems, additional functionality is needed to take full advantage of text mining. IBM Intelligent Miner for Text (www.ibm.com) includes summarization, document classification, and clustering tools. Oracle Intermedia Text (www.oracle.com) provides thematic indexing and summarization right in the database. Specialty tools, such as Megaputer's Text Analyst (www.megaputer.com) provides text mining functionality through COM objects for custom-built applications. Semio's (www.semio.co) taxonomy-generation tool can be used to automate the creation of ontologies, and Mohomine's (www.mohomine.com) tool suite includes Web crawlers and document classifiers. Of course, it is the end user's needs that will ultimately drive the set of tools required for a particular application.

TEXT: THE NEXT DIMENSION

If business information were an iceberg, text would be the bulk of the glacial object hidden below the surface and usually forgotten. Fortunately, things are changing. Commercial-quality text mining tools are available, and database vendors are recognizing the need to manage text along with numeric data. The Internet provides a wealth of raw material to complement internal documents. Whether a user needs to understand why an anomalous pattern is showing up in the data warehouse, monitor market conditions, or conduct competitive intelligence research, text is central to meeting those business intelligence needs. The time has come to accommodate documents within the workhorse of business intelligence – the data warehouse.

For critical issues facing private and public sector executives today, data mining makes the difference. From customer relationship management to risk management to improved production on the factory floor to detecting fraud and abuse, more leading-edge organizations are discovering every day that data mining gives them the ability to proactively make changes to meet and exceed their goals.

In the past few years, many businesses and public sector agencies have invested heavily in some combination of enterprise resource planning (ERP), supply chain management, sales force automation, data warehouse, and reporting software. Wanting a better return on those investments, they are now wondering whether to invest in data mining.

On the surface, data mining seems to be a risky investment. Data-mining jargon is thick, the math behind the scenes is mysterious, and data mining seems to touch only a few people in the organization.

However, when you look a little closer, the risk isn't great. In fact, it is probably much more risky not to take the plunge into data mining. The investments in ERP, data warehousing, reporting, and so on, were important but lacked leverage for three key reasons:

- They primarily replaced and updated existing ways of getting things done, which means their value was only a marginal improvement in productivity.

- Because so many organizations implemented them in more or less the same time period, they didn't offer a competitive advantage.

- Because they deal only with past events, they cannot effectively predict changes or outcomes.

We are still in the early part of the curve of the inevitable widespread use of data mining. Today, industry leaders, innovative start-ups, and progressive-thinking agencies are using data mining successfully to predict and change their futures. They are realizing not only a return on their investments in data mining, but a return on their other IT investments, especially their data warehouses.

Data mining is really about achieving your organization's goals, not about the math and the statistics. Data mining enables you to go beyond reporting and OLAP to learn not only what happened in your operations, but also why things happened. The results of data mining can easily be deployed to all the decision makers in your organization, including virtual decision makers, such as your Web site and operational systems to improve decisions in real time.

The remainder of this chapter passes along some data-mining strategies. You'll see that you don't have to be Einstein to do data mining and that data mining can have a widespread positive impact in your organization.

Don't Wait To Get Started — The Competition Is Only a Mouse Click Away

Data mining is a journey, not a project. When you've addressed today's critical issues, new ones pop up due to changes in your market or the technology. It's very likely that your competitors are already using data mining to better attract, cross-sell, and retain customers. To start before it is too late – before your reports reveal you've lost key customers – you may need to outsource your initial data mining effort.

The critical questions are

- Is your staff skilled and experienced? Most organizations do not have many people on staff, in either line of business or IT roles, with much data-mining experience. If your CIO tells you that the staff is experienced in data mining because they've built a data warehouse or have implemented OLAP, you know you've met the chief "I-don't-really-get-it" officer. You need to outsource to get the data-mining expertise you need to be successful.

- Do you have the technology infrastructure? Data mining requires data. Do you have a clean, accessible marketing database or a data warehouse? If not, to get going quickly, you can outsource this activity as well.

As you start, plan ahead for growth. As you make choices about vendors, technology, and so on, be sure to always consider scalability (the ability to work with very large data sets) and flexibility (the ability to apply the technology to a variety of situations).

Begin with the End in Mind

Personal productivity guru Steven Covey's maxim applies to your data-mining efforts. Don't be a hammer looking for a nail – there's no point in crunching a bunch of numbers or even gathering data in a data warehouse without first deciding what results you want to achieve.

The context for data mining is the issues critical for your organization's success. Start by tackling a project that is clearly linked to what you want to accomplish. Successful data mining initiatives typically start small, focusing on a critical organizational issue, such as retaining customers longer. Review your organization's strategic and business plans. Is there an area in which you aren't making the hoped-for progress? Data mining can help you get things back on track.

Decision makers tend to value data mining most favorably when they can take action based on the results. Typically, in organizations in which data mining has really taken root, the first data-mining project informed decision makers on an important topic and had both a short timeframe and clear deliverables.

Beginning with the end in mind includes defining measures that can drive improvement, delivering measures on which people are prepared to act, ensuring that the measures are easy to communicate and understand, and building measures that really fit the problem.

Focus on the "I," Not the "T"

Successful data mining project leaders retain a laser-like focus throughout the project. They focus on the people who will use the results of the project to make important decisions, the decision makers. Starting in the planning phase, keep the emphasis on delivering actionable results, not on data storage or the techniques you'll use to generate the results.

To be successful, both the decision makers (the line of business or public sector executives) and the IT organization must buy into the project plan. In most successful data mining projects, the decision maker is both the champion for and the leader of the project.

It is important to keep the decision makers involved, even in the portions of the project led by IT. Too often, decision makers and IT talk past each other and don't discover the disconnection until effort has been wasted. Avoid these disconnects by overcommunicating during all parts of the process, particularly on the first few projects. Another important part of focusing on the decision maker is to set and manage expectations. Remember that the project should have clear, timed deliverables. Build a speedboat, not a battleship.

Unless There's a Method, There's Madness

Despite the promises of early data mining evangelists, data mining is not a silver bullet for decision making; you can't just push a button and expect useful results to appear. Successful data mining projects typically use a formal and iterative process that guides the team step-by-step from selecting a critical issue through the deployment of results. Fortunately, a tested, industry-standard approach to data mining projects exists. The cross-industry standard process for data mining (CRISP-DM) was developed with the cooperation of more than 100 companies including a mix of industry leaders, small consulting firms, and academics.

There's a Reason It's Called Data Mining

Gold mining is a process for sifting through lots of ore to find valuable nuggets. Data mining is a process for sifting through lots of data to find information useful for decision making. If no gold is in a particular mountain or stream, even the best gold miner won't strike it rich. Similarly, the data itself is critically important to data miners.

After you've picked a problem to solve, you can start thinking about data. The data is so important that three of the 10 points in this section focus on data. Start at the highest level; think about the data that you need to gather from the perspective of the information you want to deliver. You'll want to capture the detail of data needed for everything from strategic analysis and tactical alerts.

Two high-level tips from successful data miners about data are

- Most often, the unit of analysis in a data mining project is a customer or constituent. In order to do analysis at that level, you probably need a unique ID by customer or constituent everywhere that you capture and store data.

- Make use of metadata – information about your data – wherever you can. Metadata includes simple information, such as the source of the data, to more advanced information, such as when the data was last changed or whether the data has been imputed (a missing part of a record has been filled in based on information in other parts of the record).

Better Data Means Better Results

Better data is only one of a handful of ways you can improve your data mining results. Better data means comprehensive and more accurate analysis. Typically, it's far more valuable to include more variables or columns of data in the analytical process than it is to have more cases or records. However, you encounter a trade-off between using many variables and getting useful results quickly. Because data mining is a journey, successful data miners typically work with the data they have, get results, realize some ROI, and then add data over time to become even more effective.

The best analysis is done using three types of data:

Transaction data. Transaction data is very powerful; it tells you what the customer has actually done. Psychologists have proven that past behavior is the best predictor of future behavior. The good news is that most organizations have a great deal of transaction data – everything from prior purchases or donations to a record of which Web pages a person visited and how long he or she lingered on them.

Purchased data. When customers or constituents interact with you, they typically tell you only a subset of useful information. An entire industry exists to provide very useful supplemental data about the customer's current situation, including demographic and psychographic data.

Collected data. Collected data offers a great opportunity for leverage in data mining. And, because it takes effort and skill to collect useful data, collected data offers a unique opportunity for competitive advantage. Collected data adds information about a customer's attitudes and opinions into the analysis phase and results in better decision-making information. You can collect data about customer satisfaction levels, customer preferences, purchase intentions, share-of-wallet information, and so on. Or, you can turn to professionals in the market research industry to collect it for you.

Although reporting and OLAP are informative about past facts, only data mining can help you predict the future of your business. Table 10.3 compares the different questions posed by OLAP and data mining.

It's Still Garbage in, Garbage out

Some things never change. In the early days of computing, the phrase *garbage in, garbage out* was coined to reflect the reality that computing results are dramatically affected by the quality of data. Getting access to the right data, cleansing it, and preparing it for analysis is typically the most time-consuming step of the data mining process. Don't fool yourself – 70 to 80 percent of time invested in data mining projects is typically used for data access, cleansing, and preparation. So, plan and set expectations accordingly.

This work is often time-consuming and many people do not find it particularly exciting. In addition, dramatic increases in the usefulness of the end results can proceed from more analytically sophisticated approaches to data preparation. As a result, this step is often a good candidate for outsourcing.

Table 10.3.

Data Mining Helps Predict the Future

OLAP	Data Mining
What was the response rate to our mailing?	What is the profile of people who are likely to respond to future mailings?
How many units of our new product did we sell to our existing customers?	Which existing customers are likely to buy our next new product?
Who were my 10 best customers last year?	Which 10 customers offer me the greatest profit potential?
Which customers didn't renew their policies last month?	Which customers are likely to switch to the competition in the next six months?
Which customers defaulted on their loans?	Is this customer likely to be a good credit risk?
What were sales by region last quarter?	What are expected sales by region next year?
What percentage of the parts we produced yesterday are defective?	What can I do to improve throughput and reduce scrap?

Avoid the OLAP Trap

Many vendors try to tell you they have all you need for data mining or effective online personalization. That's unlikely to be true. Successful data mining requires three families of analytical capabilities: reporting, classification, and forecasting.

Why are all three types needed? Each capability delivers a different kind of information. Reports inform management of what has happened in the past. Reports (including OLAP) are popular because they are easy to understand and easy for IT to produce. However, if all you have is reporting, you can see only what has already taken place. You cannot predict what might happen later. It's as if you're trying to drive your car by looking only in the rearview mirror; the faster you need to go, the more risky it is.

Classification and forecasting are different from reporting because they enable you to gain more understanding about why things happen and to make predictions about what is likely to happen under different scenarios. Armed with this information, you can make proactive changes in your organization and realize better results. Together, classification and forecasting are commonly known as *predictive modeling*. Classification methods put things into groups; for example, customers likely to spend more with your company or constituents likely to vote for a proposition. Typically, the classification process includes two steps: establishing the groups and determining (or predicting) group membership on a case-by-case basis. Forecasting methods deal with data where time is the critical measure. Examples of forecasting are sales by product and/or region over time and population growth over time.

It's not the purpose of this article to delve into the pros and cons of the various methods of predictive modeling. The key point here is contrary to what is often heard: algorithms do matter. Algorithms matter because, in the end, just any result won't do; to compete and win, you must have the best answer. Getting to the best answer involves powerful and comprehensive solutions that enable you to determine preferences of not only current customers, but also those for whom you have no purchase history. Just as a carpenter uses more than a hammer to build a house, a data miner uses more than one analytical method to get the best results.

Experienced data miners offer three suggestions with regard to data mining analytics:

 Don't redo your existing reports without adding value. Typically, that means adding one of these three things: drill down, alternative views of the same information, and prediction.

● Always offer ad hoc capabilities. Canned reports are a great starting place, but they are just a starting place. In today's fast-paced world, ad hoc capabilities are a requirement.

● Your data mining software should offer an easy way for you to incorporate your business knowledge. Software that doesn't enable you to use what you know about your business when building a model is simply not going to give you the best results.

Deployment Is the Key to Data Mining ROI

The ultimate goal of data warehouse ROI cannot be achieved without data mining, but truly successful data mining cannot be achieved without deployment. Deployment means getting the information, in a usable format, to the place where it is needed.

The four types of deployment are as follows:

To decision makers. Getting information into the hands of people who can effect change is key. Typically, deployment to decision makers is done via intranets. Two ways people can go one better in this regard are offering live deployment (in which the decision maker can interact with the results) and supporting offline use of the information for travelers.

To virtual decision makers. If a customer enters your store, he or she can receive personalized service from one of your employees. If a person comes to your Web store, you can still offer a personalized experience if you deploy predictive models. For example, based on a combination of what you know about a customer from historical transactions and purchased data and his or her actions at your Web site, your virtual decision maker can instantly offer different Web page content, alternative products, unique discounts, and so on.

To operational systems. Another customer touchpoint is your call center. By prompting your call center representatives, deployment can enable the same type of personalization as in the Web scenario previously mentioned. In a manufacturing setting, a deployed model may take information coming from a production line and, based on that data, either send a message to a troubleshooter or even make adjustments without human intervention.

To databases. Interactions with customers today are many faceted. For example, a retailer may interact with the same customer via a storefront, the Web, a call center, and a catalog. In order to keep information about that customer current, savvy organizations store all data in a centralized data warehouse. When a customer's profile is updated, the *score*, which indicates the customer's category, is deployed back to the database for use in future interactions. For example, recent purchase activities or updated demographic information cause a classification model to change the category into which a customer is classified.

Champions Train So They Can Win the Race

Most corporations today have a handful of heavy-duty model builders (data miners), a fair number of knowledge workers (analysts), and lots of information consumers. In addition to the obvious returns from training the IT personnel and data mining software users, you are likely to see greater overall return from your data mining investment if you also educate the people who receive the information and use it to make decisions. Although not everyone wants to be an analyst or a model builder, ongoing success in data mining typically requires some education and training for your information consumers.

The type of education and training needed varies by the individual's role in the process. Many people have paid their analytical dues in college by sitting through a statistics course, and volunteers to repeat that are few and far between. However, many benefit from a more practical refresher course that emphasizes analytical thinking, not analytical methods. The content can be delivered in a variety of ways, from a classroom setting to just-in-time computer-based tutorials.

Another key training consideration arises if you outsource your first data mining project to get started quickly. If you do, you probably want to train your IT staff to be ready to manage the system, make updates, and so on, when you assume responsibility for it.

MINE YOUR DATA

Today, data mining can make the difference in every industry and organization throughout the world. You can mine your data and use the results to determine not only what your customers want, but to also predict what they will do.

During the last several years, data mining techniques have been used by companies to understand the demographics of their customers and to provide them with personalized interactions. Various data mining techniques have been deployed in order to identify hidden trends and new opportunities within the data. These various data mining techniques have been embedded into software applications that process complex algorithms in order to provide meaningful information. Although end-user data mining applications are available, they have not been extensively deployed throughout organizations because they are often not understood. One way to understand the capabilities of data mining is to compare it to other business intelligence (BI) technologies.

CHAPTER 11

Ad Hoc Queries

Topics in this chapter:

- Online Analytical Processing (OLAP)

- Data Mining

- Integrating Data Mining into the Decision-Making Process

With an ad hoc query application, users have the ability to access information on demand. What they ask for is what they get. For example, a user creates and executes an ad hoc query that answers the question, "How much revenue was generated by each customer during this year?" The results from the query contain customer name and revenue for the year selected. Figure 11.1 is a representation of the result set produced by the ad hoc query.

The revenue by customer could be totaled to another question, "How much revenue was generated this year?" In addition, other questions, such as "What customer generated the most revenue for the company?" and "What customer generated the least amount of revenue for the company?" could also be answered. Although the original query result was useful and addressed several questions, this business intelligence (BI) technology will not identify unusual patterns or reveal unusual relationships. What the user

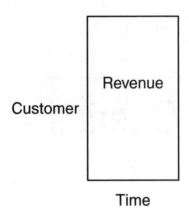

FIGURE 11.1.

The result set produced by the ad hoc query, "How much revenue was generated by each customer during this year?"

requested was revenue by customer for the current year and that is the information provided – no more, no less.

ONLINE ANALYTICAL PROCESSING (OLAP)

OLAP applications provide users with the ability to manually explore and analyze summary and detailed information. For example, a user creates and performs an OLAP analysis that answers the question, "What was the revenue for each quarter of this year by geographic region and customer?" The results from this analysis would contain geographic region, customer name, revenue, and quarters selected. Figure 11.2 is a representation of the result set produced by the OLAP analysis.

Additional questions could be posed of the data, which could highlight seasonal revenue patterns by geographic region. However, a user who understands how to navigate the data must direct this process. OLAP can highlight only the patterns within the data requested. The user must identify

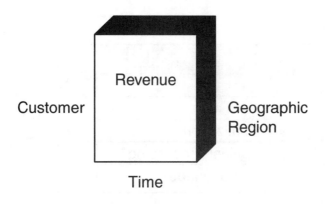

FIGURE 11.2.

The result set produced by the OLAP question, "What was the revenue for each quarter of this year by geographic region and customer?"

the trends and patterns highlighted by the OLAP analysis. This BI technology will not identify unusual or reveal hidden relationships.

DATA MINING

Data mining can best be described as a BI technology that has various techniques to extract comprehensible, hidden, and useful information from a population of data. Data mining makes it possible to discover hidden trends and patterns in large amounts of data. The output of a data mining exercise can take the form of patterns, trends, or rules that are implicit in the data.

Various data mining techniques can be deployed, each serving a specific purpose and varying amounts of user involvement. Figure 11.3 displays the progression of data-mining techniques in the order of user involvement.

Neural networks are highly evolved systems that provide predictive modeling. These systems are very complex, and it takes time to train the system

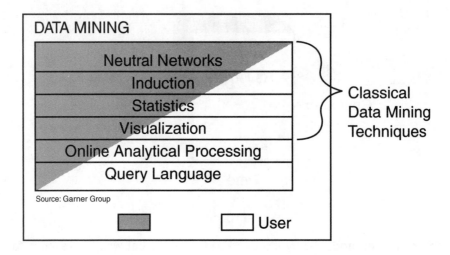

FIGURE 11.3.

The progression of data-mining techniques

to perform human-like thinking. This data-mining technique has been used to detect potential fraudulent credit card transactions.

Induction is a data-mining technique that induces rules inherent within the data. The rules are used to understand the relationships that exist. A classic example is, *When people buy diapers, they also buy beer 50 percent of the time.*

Statistics is the basis of all data-mining techniques and requires individuals who are highly skilled in mathematics to build and interpret the results.

Visualization displays the data in a graphical or three-dimensional map, thereby enabling the user to identify trends, patterns, and relationships. Although an image that is produced provides another perspective of data relationships, visualization is often incorporated into data-mining applications.

Although *OLAP* and *query language* are listed by the GartnerGroup as data-mining techniques, the amount of user involvement is extensive and extremely time-consuming to identify hidden trends and relationships. Therefore, using such techniques is not cost-effective.

Using a data-mining application, a user can ask, "What are the distinguishing characteristics of our credit customers who pay on time?" The results from the data-mining exercise would then be used to create the condition statement of an ad hoc query, which identifies customer names and contact information within the database for the purposes of cross-selling additional services.

Ad hoc query applications scratch the surface of the value that exists within a database, and OLAP provides users with greater depth and understanding. However, data mining digs deeper and provides users with knowledge through the discovery of hidden trends and relationships. The combination of data mining with an ad hoc query or OLAP application is extremely powerful and provides users with knowledge about the data being analyzed and the ability to act upon the knowledge. Figure 11.4 depicts the value and purpose of the BI technologies.

Data mining can play a key role in developing a business strategy and is a fundamental and compelling factor for justifying, designing, and building intelligent data warehouses. The transition from data mining buzz to practical implementation, however, has been slow to come, although it is quickly gaining momentum. Many reasons are evident as to why it has taken time for the technology to take off. First, this technology is extremely powerful, and companies are reticent to adopt a technology without the core competency to use it. Second, there is an enormous cultural shift that comes with automating and integrating application intelligence into the decision-making process. Third, the technology has been somewhat raw and geared toward analytics researchers and experts, not necessarily toward mainstream business people.

In the last few years, companies have been quietly adopting data mining with success. What is the essential difference that has caused the acceptance and usage of the technology? First, vendors have become proactive in teaming up with universities and consulting groups to provide the massive education required to bring business and IT up to speed. Second, vendors are tuning into the natural marriage of warehousing and mining and are developing more robust architectures that take advantage of warehousing as

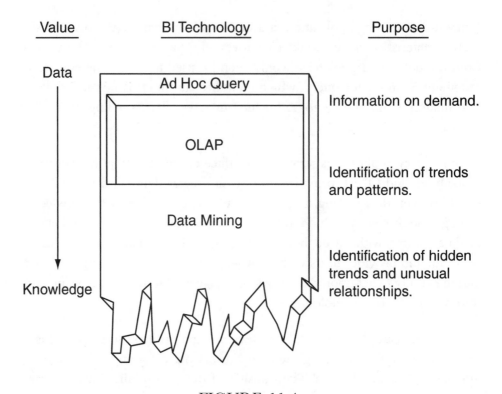

Value	BI Technology	Purpose

FIGURE 11.4.

The value and purpose of BI technologies.

the enabling infrastructure for enterprise mining. Third, both vendors and consulting groups have been packaging data mining into software and processes, respectively, enshrouding the raw horsepower with sleek, application-oriented, user-friendly systems. Fourth, the competitive urgency is overshadowing the tendency to watch and wait.

With company mission statements focused on customer-centric objectives, organizing and managing processes from the customer perspective is becoming critical. Data mining is a powerful ally in aligning business strategies to acquiring, retaining, and growing customers. The best asset available to organizations in realigning business processes to the customer perspective is that the internal historical data integrated with external demographic and psychographic data. Building warehouses with mining at the heart brings a streamlined focus to design and content and delivers immediate exploitative capability to warehouse users.

A customer-centric warehouse can be mined and analyzed by taking information about consumers and customers that is derived from point-of-service (POS) transactions and primary research. For example

- Information about the public at large is mined to create segments around which businesses position their products and services.

- Key buying behavior and physical and lifestyle attributes of existing customers are used to learn a more effective means of targeting and acquiring new customers.

- The same information is used to spot potential customer defectors before they leave.

- Individual transactions are analyzed to spot product assortments, yielding a better understanding of what products customers buy in groups.

- The same information is used to design product bundles and to cross-sell strategies that will ensure customer loyalty through time.

- Analysis is performed to understand profitability segments so that promotions and services can be targeted to those that are most profitable.

Further analysis feeds everything from promotion planning to shelf-space management. Data mining supplies information for developing strategies that foster customer growth and satisfaction and discovers those factors that predict bottom-line profitability.

In today's world, allocating capital and expense is grounded in experience and an in-depth knowledge of functional operations. Processes and methods for forecasting and allocation have been proven time and again. For the most part, companies make these decisions from a well-informed basis and have a solid analytical footing to judge where the investments should be allocated. The marketing function, however, is often an exception.

A culture clash is taking place between traditional marketers and database marketers, which centers around the use of data mining. Whereas traditional marketers rely on mass communication campaigns geared at acquiring new customers, database marketers have embedded data-mining techniques into their analysis processes, squeezing the most out of the current customer base. Database marketers clearly have more proof-of-performance numbers on their side. They are proficient at using data mining to make management comfortable by providing direct financial and economic return results from their promotional campaigns. The end result is that decision making within the marketing discipline is split between informed and uninformed resource allocation decisions.

The irony is that the answer to providing the right metrics to judge the performance of all marketing functions lies within the existing marketing database. However, many companies never bother to look. Database marketers are accustomed to embedding data-mining tools directly into their business processes; yet the remainder of the organization chooses not to. A clear competitive advantage exists for companies willing to make this change.

It should be clear by now that data mining is a business function and can provide a strategic advantage in developing, defining, and deploying competitive business strategies. Consider two areas when successfully ushering data mining into an information environment: skill sets and technology.

INTEGRATING DATA MINING INTO THE DECISION-MAKING PROCESS

Skill sets will vary by the data-mining stakeholders in your organization. The skill sets for each of the stakeholders include:

Stakeholder	Skill Set
Miner	Analytics, model building, statistics, neural net development, research
Domain expert	Intensive business and data knowledge, experience, decision maker
Business user	Understanding of business and data, decision maker, user of mining results
IT	Support of analytic environment, data model for new DM components, integration of DM (tools, processes, results, and models) into DW

Technology integration points include communication links between data-mining software and both data and application domains. Data links include sourcing, transformation, and loading of input and output variables and result sets. Application links include the capability to invoke data-mining tools from the DW environment; access to mining results and visualizations; and the capability to invoke analytic models for prediction and description from the operational and decision-support systems, both back end and front end.

Data mining has evolved from manual statistical methods to desktop mining to enterprise mining. With appropriate skill sets, the right team, a warehousing infrastructure, and data-mining tools, companies can transition into agile competitors who maneuver quickly with the global demands of the marketplace.

CHAPTER 12

Data Warehousing on the Web

Topics in this chapter:

- Farming Web Resources for the Data Warehouse
- Using the Web for Business Intelligence
- Reliability of Web Content
- Information Flow
- Where Are We Heading?

FARMING WEB RESOURCES FOR THE DATA WAREHOUSE

The Web has become many things to many people. To those in the data warehousing profession, the Web has been irrelevant or maybe a threat. As a chaotic and unmanageable influence, the Web can be perceived as a threat to the security and tranquility of the warehouse environment.

Although Web technology is used extensively as the delivery mechanism for warehouse data, no one has seriously considered using Web content as input to the data warehouse. The paradigm of the Web is radically different than that of the data warehouse. Adapting an old programming term, one might say that Web content is *spaghetti data* (lots of links with little discipline). Web content is highly volatile and diverse, challenging our imagination to discover those nuggets having real business value.

In many ways, the Web is the mother of all data warehouses. The Web is becoming the universal delivery mechanism for global data. However, the immense information resources of the Web are largely untapped by data warehousing systems.

USING THE WEB FOR BUSINESS INTELLIGENCE

Professor Peter Drucker, the senior guru of management practice, has admonished IT executives to look outside their enterprises for information. He remarked that the single biggest challenge is to organize outside data because change occurs from the outside. He predicted that the obsession with internal data would lead to being blindsided by external forces.

The majority of data warehousing efforts result in an enterprise focusing inward; however, the enterprise should be keenly alert to its externalities. As markets become turbulent, an enterprise must know more about its customers, suppliers, competitors, government agencies, and many other external factors. The information from internal systems must be enhanced with external information. The synergism of the combination creates the greatest business benefits.

RELIABILITY OF WEB CONTENT

Many question the reliability of Web content, as they should. However, few analyze the reliability issue to any depth. The Web is a global bulletin board

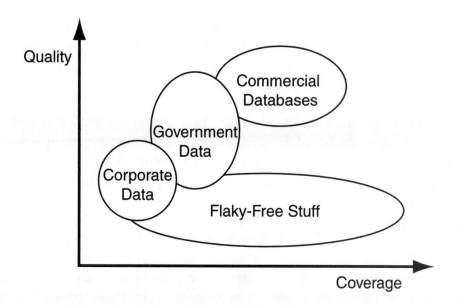

FIGURE 12.1.

Web-based information resources

on which both the wise and the foolish have equal space. Acquiring content from the Web should not reflect positively or negatively on its quality.

Consider the following situation: If you hear, "Buy IBM stock because it will double over the next month," your reaction should depend on who made that statement and in what context. Was it a random conversation overheard on the subway, a chat with a friend over dinner, or a phone call from a trusted financial advisor? The context should also be considered when judging the reliability of Web content.

Think of Web resources in terms of quality and coverage, as shown in Figure 12.1.

Toward the top are information resources of high quality (accuracy, currency, and validity), and resources toward the right have a wide coverage (scope, variety, and diversity). The interesting aspect of the Web is that information resources occupy all quadrants.

In the upper center, the commercial online database vendors traditionally have supplied businesses with high-quality information about numerous topics. However, the complexity of using these services and the infrequent update cycles have limited their usefulness.

More to the left, governmental databases have become tremendously useful in recent years. Public information was often available only by spending many hours of manual labor at libraries or government offices. The Electronic Data Gathering, Analysis, and Retrieval (EDGAR) database maintained by the U.S. Securities and Exchange Commission contains extensive information on publicly traded companies and is updated daily.

At the left, corporate Web sites often contain vast amounts of useful information in white papers, product demos, and press releases, eliminating the necessity to attend trade exhibits to learn the "latest and greatest" in a marketplace.

Finally, the "flaky-free" content occupies the lower half of the figure. Its value is not in the quality of any specific item but in its constantly changing diversity. Combined with the other Web resources, the flaky-free content acts as a wide-angle lens to avoid tunnel vision of the marketplace.

INFORMATION FLOW

The data warehouse occupies a central position in the information flow of a Web farming system, as shown in Figure 12.2.

Like operational systems, the Web farming system provides input to the data warehouse. The result is to disseminate the refined information about specific business subjects to the enterprise.

The primary source of content for the Web farming system is the Web because of its external perspectives on the business of the enterprise. As a content source, the Web can be supplemented (but not replaced) by the intranet web of the enterprise. This content is typically in the format of

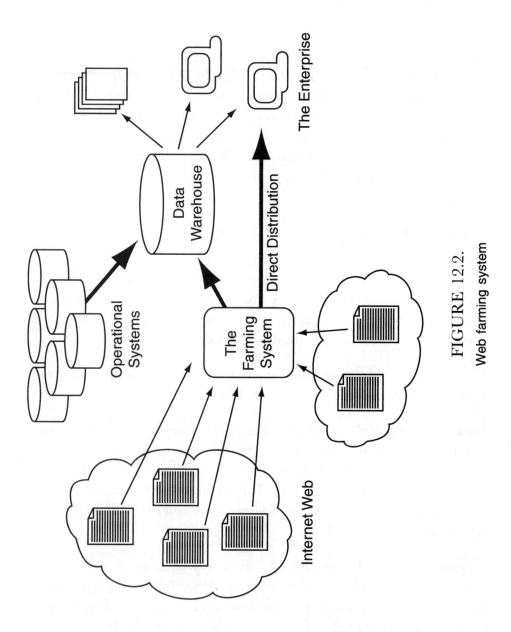

FIGURE 12.2.
Web farming system

internal Web sites, word processing documents, spreadsheets, and e-mail messages. However, the content from the intranet is usually limited to internal information about the enterprise, thus negating an important aspect of Web farming.

Most information acquired by the Web farming system will not be in a form suitable for the data warehouse. It will either be unstructured hypertext or unverified tabular values. In either case, you must refine that information before loading it into the warehouse. Even in this unrefined state, this information could be highly valuable to the enterprise. The capability to directly disseminate this information may be required via textual message alerts or "What's New" bulletins.

Refining Information

When a data warehouse is first implemented within an enterprise, a detailed analysis and reengineering of data from operational systems is required. The same is true for Web farming. Before Web content can be loaded into a warehouse, the information must be refined.

Four processes for refining information exist: discovery, acquisition, structuring, and dissemination, as shown in Figure 12.3.

Discovery is the exploration of available Web resources to find those items that relate to specific topics. Discovery involves considerable detective work far beyond searching generic directory services, such as Yahoo!, or indexing services, such as Alta Vista. Further, the discovery activity must be a continuous process because data sources are continually appearing – and disappearing – from the Web.

Acquisition is the collection and maintenance of content identified by its source. The main goal of acquisition is to maintain the historical context so that you can analyze content in the context of its past. A mechanism to efficiently use human judgment in the validation of content is another key requirement.

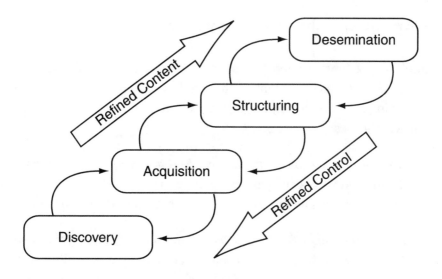

FIGURE 12.3.

The four processes for refining information

Structuring is the analysis and transformation of content into a more useful format and into a more meaningful structure. The formats can be Web pages, spreadsheets, word processing documents, and database tables. As we move toward loading data into a warehouse, the structures must be compatible with the star-schema design and with key identifier values.

Dissemination is the packaging and delivery of information to the appropriate consumers, either directly or through a data warehouse. A range of dissemination mechanisms is required, from predetermined schedules to ad hoc queries. Newer technologies, such as information brokering and preference matching, may be desirable.

A continuous flow exists among the processes, rather than a step-wise procedure. Further, a bidirectional flow also exists among the processes. The left-to-right flow refines the content of information, which becomes more structured and validated. The right-to-left flow refines the control of the processes, which becomes more selective and discriminating.

Rendezvous with the Data Warehouse

Assume that we have refined (discovered, acquired, and structured) some collection of Web-based content so that we have confidence in its validity. The following are various ways in which we can integrate Web content into current warehouses:

1. *Augment the descriptive information about a dimension.* For example, a mailing address can generate a longitude-latitude coordinate that can retrieve a satellite image of the area around that location.

2. *Add nominal (or ordinal) data about a dimension so that more options for pivoting cross-tabulations are available.* For example, a mailing address can generate a longitude-latitude coordinate that can classify a customer into a sales region.

3. *Add interval (or ratio) data about a dimension so that correlation analysis with other dimensions can be performed.* For example, a mailing address can generate a census tract ID, which can give family income, household size, population density, age distribution, and so on. Or, extensive financial data about publicly traded corporations can be retrieved from the SEC databases.

4. *Create a new dimension table.* For example, recording daily weather as an additional dimension for analyzing sales patterns.

5. *Create a new fact table based on an external event.* For example, a count of the number of mentions your product receives versus the number of mentions that the competitors' products receive over weekly intervals within a set of trade publications.

Consider Table 12.1, which lists examples of star-schemas for portions of an enterprise data warehouse.

These examples show some of the ways in which data, external from the company, can enhance the value of these data marts.

Table 12.1.

Star-Schemas for Different Portions
of an Enterprise Data Warehouse

Mart Schema	Fact Table	Dimension Tables
Retailing	Sale	Time, Product, Store, Promotion
Inventory	Item	Time, Product, Supplier, Location
Shipments	Shipment	Time, Product, Customer, Deal, Ship-From, Ship Mode
Banking	Account	Time, Product, Branch, Household, Status
Subscriptions	Subscription	Time, Product, Customer, Status, Promotion
Insurance	Policy	Time, Policy Type, Agent, Coverage, Client, Status
Hospital Procedure	Procedure	Time, Procedure Type, Patient, Hospital, Physician, Assistant, Diagnosis
Frequent Flier	Trip	Time, Customer, Flight, Airports, Fare Class, Sales Channel, Status
Hotel Stays	Visit	Time, Customer, Hotel, Sales Channel, Status

First, all schemas have a time dimension for the period during which the fact occurred. Other valuable external information that could be added includes critical events (economic, political, military, and so on) that happened during that period, the actual weather and weather predictions, holidays and seasonal changes, and other events that could affect the flow of commerce. Businesses are not isolated from the effects of natural, social, political, and economic events occurring throughout the world.

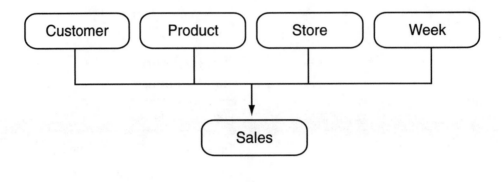

FIGURE 12.4.

Sales warehouse data schema

Second, most schemas have a product dimension that contains attributes about the company's offerings. Valuable external information to consider includes recent mentions in the trade press, counting mentions within a set of trade press by week and correlating those mentions to sales, links to competitors' products for comparison, and the prices for the same product through alternative distribution channels.

Consider a simple data schema for a sales warehouse as shown in Figure 12.4. In this warehouse, we have sales data by customer, product, and store, aggregated on a weekly basis. Let's assume that we have mostly corporate customers, rather than individuals, as in a large office furniture company.

Web farming would be valuable by enhancing the demographics (for example, quarterly financials) about customers. As shown in Figure 12.5, by adding information on customer demographics, selective marketing can be performed based upon the profitability and requirements of customers. By knowing what types of customers buy what types of products at which stores, we can promote specific sales and anticipate demand. For example, companies that are expanding are more likely to order office furniture.

The demographic information is added to the customer dimension so that analyses on specific customers are enhanced by the demographics, as shown in Figure 12.6. As experience with the demographics matures, data-mining

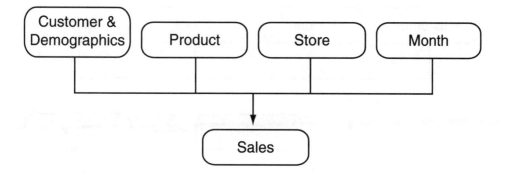

FIGURE 12.5.

Addition of customer demographics.

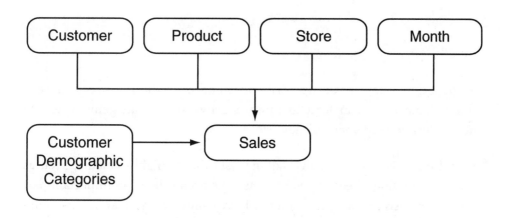

FIGURE 12.6.

Demographics as a separate analysis dimension

techniques can cluster customers into segments based on demographics. Then demographics can be meaningfully categorized and can become a separate analysis dimension.

Another example of using Web farming to enhance a data warehouse is the addition of demographics on the store, as shown in Figure 12.7. Census data about the communities surrounding the store can be added as another

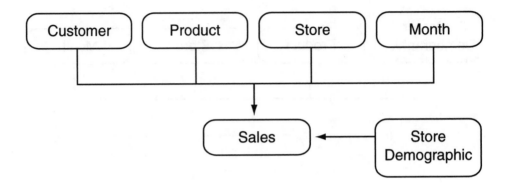

FIGURE 12.7.

Addition of store demographics

business dimension if you use ZIP codes and even the full street address. This enhancement can lead to more effective store management and to more effective placement of new stores.

A final example could involve adding data that is highly volatile, such as weather conditions. Seasonal variations have always been an important part of sales analysis. However, a sudden heavy snowstorm or an intense hailstorm can also affect the sales of specific products, in addition to the seasonal variations. This example shows that timely and continuous flow of Web content into the warehouse can aid in the day-to-day management of the business.

WHERE ARE WE HEADING?

In many ways, the data warehouse is not a requirement for Web farming. You could successfully farm the Web, bypassing the data warehouse, and still reap value for the enterprise. However, this approach may achieve short-term gain at the expense of long-term potential. For Web farming to be successful in the long term, it should be integrated into the enterprise data warehouse.

Across the industry, the current practice of data warehousing is fulfilling its promises of delivering real business benefits. With Web farming, we are challenged with deeper issues concerning information refinement and knowledge management. Web farming will force change upon the data warehouse. However, this change will evolve the data warehouse into a better system of knowledge management for the enterprise.

CHAPTER 13

OLAP versus Data Mining: Which One Is Right for Your Data Warehouse?

Topics in this chapter:

- The Data Warehouse Model

- Data Warehouse Model

- Summary

The final piece of your data warehouse architecture is a top-notch data analysis tool to access and slice and dice the data. These tools fall into two broad categories: OLAP (online analytical processing) and data-mining tools. Many tool vendors hawk any analysis tool they sell as a data-mining tool, and sometimes the market seems more like a carnival midway – with unfamiliar people screeching at you to take a peek and purchase what you like – than it does a business market. However, to get the best analysis tool, and type, for your data warehouse, you have to look past the hype and understand what each type of tool is and isn't. That's not always easy, and you may end up needing a hybrid that combines the best of both worlds, but in order to gain maximum productivity from your data warehouse, you must take a stroll down the midway and look inside each booth.

As I said, OLAP and data-mining tools are often lumped in the same category. In reality, they are two different animals. OLAP tools are the workhorses of the analysis tool family. They provide a range of functionality that spans standard reports and queries to multidimensional analysis capability. For example, you might say, "I want to see how many widgets we sold last year, sorted by type." That's a simple query that produces a simple report. OLAP tools also can go further. They can slice and dice data by certain regions, business units, categories, and time periods – for example, by multiple dimensions. Your multidimensional query might go something like this, "I want to see how many widgets we sold by type, then by business unit, in the second quarter of the year, by region, comparing 1998 to 1999."

Some examples of OLAP tools include AlphaBlox from AlphaBlox Corporation, BusinessObjects from BusinessObjects SA, PowerPlay by Cognos, MicroStrategy7 from MicroStrategy, Inc., and Essbase from Hyperion Solutions (formerly Arbor Software).

Data-mining tools are racehorses and can go even further (sort of). They can provide you with information you didn't even know you were looking for. That's the beauty and the beast of data-mining tools. You can really mine business-data gold with them, but you can also miss the gold and dig up a lump of coal instead if you're not careful. The two basic uses for data-mining tools are modeling and segmentation and what Jill Dyché, author of *e-Data: Turning Data into Information with Data Warehousing*, calls *knowledge discovery*.

This term really describes what data-mining tools enable you to accomplish. They enable you to take mountains of raw data from all channels, coalesce and analyze it, and look for trends and patterns in customer habits, so that you can discover more about your customers and what they want from you. Some examples of data-mining tools include DataSage from Vignette Corporation (formerly from DataSage, Inc.), Enterprise Miner by SAS, Intelligent Miner by IBM, and Clementine by SPSS.

With good information, you can categorize your customers by different parameters, much like you categorize your products. You can put them into different, and multiple, segments according to any category you choose, such as region, income range, gender, profession, age, and so on. You also can model their behavior to predict future sales and revenues.

However, data-mining tools give you even more analysis capability. The top-notch, true data-mining tools give you powerful statistical analysis capability, which enables you to learn things about your customers that you didn't even know you wanted to know. There's an old piece of data mining lore that amply illustrates this point. One convenience-store chain used a data-mining tool to analyze the purchasing patterns of its customers. Of all things, the marketing research team found that the most highly correlated purchases, by time period, were beer and diapers on Friday evenings. Daddy or Mommy, coming home from a hard week at the office, stopped to pick up a six-pack and picked up diapers for Junior at the same time. So, to make their customers' lives easier (and increase sales) the chain started stocking beer and diapers side by side. It worked.

With all the spiffy tools available and with all the heady functionality they provide, how can you possibly know which one you need? My advice to you is that if you sell any type of tangible product or high-volume services, you need both. Any organization that sells anything that is not completely customized for each customer needs both OLAP and data-mining capabilities. You may not need the full-blown power of the best data-mining tools on the market, but any OLAP tool you purchase needs to at least provide you with baseline modeling and segmentation functionality that mimics data-mining capabilities.

With that said, for one business decision that you make, you absolutely need data-mining capabilities, and the more powerful the tool is, the better. If your organization is planning to implement a customer relationship management (CRM) initiative, you absolutely must have a true data-mining tool. Without the powerful statistical analysis capabilities that data-mining tools provide, the valuable information contained in the mountains of clickstream, and other data, that your organization collects on your customers will get more lost than the proverbial needle in a haystack.

CRM initiatives are the rage in the IT world today for one reason: They help increase profits. Happy customers equal higher revenues. Knowing this, how do happy customers and CRM initiatives fit together? If the goal is to get and keep happy customers, your business must become totally customer centric, across the organization. The mission of every business unit and every level of worker must be to satisfy the customer. To do this, you need information about your customers and lots of it. I didn't say *data*; you probably have more of that already than you know what to do with.

You can turn that data into usable information with data mining. With the analysis capability you gain by implementing these powerful tools, you can spot customer trends and patterns that you wouldn't normally detect and customize your Web site or marketing and promotions to suit those customers' needs. It is the best oxymoron going: mass customization. Your customer visits your Web site, is greeted by name, is shown products that are in favored-purchase categories, and is speedily checked out after purchasing goods. The customers feel like they are your favorite customers. They're happy, and happy customers equal increased revenues.

As with any business decision, especially those involving IT purchases and projects, the decision to use an OLAP tool, a data-mining tool, or a situationally dependent combination of both is a tricky one. You must know what you want to gain out of the tool and what you stand to lose if you make the wrong decision. In the end, it may come down to a question of your organization's direction. Do you just want to keep your customers, or do you want to keep your customers happy? If you want to keep your customers happy, you'll learn more about them. How much you want to learn will largely be the deciding factor on how powerful of a tool you choose.

THE DATA WAREHOUSE MODEL

Opinions differ concerning the structure of the data warehouse model and the structure of the data marts that are to be used directly by business people to perform multidimensional analysis. The standard approach is that the data warehouse model should be relationally founded with deviations from third-normal form, which recognize the role of the data warehouse.

Data Warehouse

To understand the reasons for the different structures, examine the roles of the data warehouse and the data marts. The data warehouse, as shown in Figure 13.1, is the target of data acquisition and the source for data delivery.

The data warehouse serves the following primary functions:

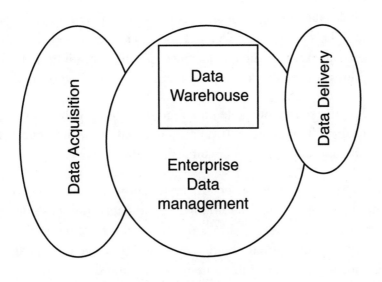

FIGURE 13.1.

Data warehouse position

- It is a reflection of the business rules of the enterprise – not just of a specific function or business unit – as they apply to strategic decision support information. This characteristic requires resiliency to easily accommodate changes to the business rules, which include new data elements, shifts in hierarchical relationships, or changes to relationships between existing entities.

- It is the collection point for the integrated, subject-oriented strategic information that is handled by the data acquisition process. This characteristic calls for a modeling technique that supports subject-orientation and provides the flexibility to integrate data from additional sources over time.

- It is the historical store of strategic information, with the history relating to either the data or its relationships. This characteristic calls for a modeling technique that easily supports incorporation of history.

- It is the source of information that is subsequently delivered to the data marts. The data marts in question may be used for exploration, data mining, managed queries, or online analytical processing. This requires the model to provide unbiased data that can subsequently be filtered to meet specific objectives. It further requires the model to support summarized and aggregated data.

- It is the source of stable data regardless of how the processes may change. This requires a data model that is not influenced by the operational processes creating the data.

The data warehouse serves a secondary function to directly support queries. Although there should be nothing in the data warehouse design that should inhibit these queries, this is not the primary function and is only a secondary consideration in the design. Queries that go directly against the data warehouse do not need to be user friendly and do not need to provide rapid response time. Some optimization of the model, such as partitioning, incorporation of additional indexes, and further denormalization, is often appropriate, because it improves query performance without detracting from the data warehouse's capability to serve its main role. Additionally, based on the technical environment, including the ETL tools, additional modifications may be appropriate.

Data Marts

The data mart established for online analytical processing, as shown in Figure 13.2, is the target of data delivery and the direct source of data accessed by end users.

The OLAP data mart serves the following primary functions:

- It is a reflection of the business rules of a specific function or business unit – not the enterprise – as they apply to strategic decision support information. The rules need to be consistent with the enterprise rules but are tailored to the business capabilities addressed by the data mart.

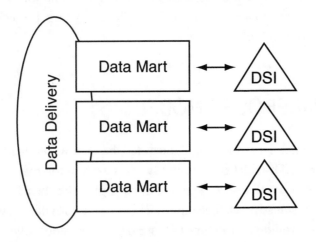

FIGURE 13.2.

Data mart position

- It obtains the data from a relatively stable, cleansed, and integrated source. This removes the requirement to handle cleansing and integration.

- It is a set of tables designed for direct access by users who need to analyze data according to a set of predefined parameters (for example, dimensions). This characteristic requires a structure that supports easy, intuitive analysis across these parameters and hierarchies within those parameters.

- It is a set of tables designed for aggregation. The fact table may contain detailed data, but most of the queries will view this data summarized by the business constraints that form the dimensions. For example, the data mart may contain sales transaction data, and the user query will be for a summary of sales transactions by a time period, product line, and store. This characteristic requires a structure that supports easy, intuitive data aggregation across appropriate parameters (for example, business constraints or dimensions).

- It is typically not a data source for traditional statistical analysis needed for exploration or data mining. Omission of this requirement permits the data to be collected along predefined, known relationships.

DATA WAREHOUSE MODEL

The fundamental data modeling choices for the data warehouse are relational or dimensional. The relational model satisfies each of the major requirements given for the data warehouse. The relational approach is best for reflecting the enterprise's business rules. This approach can be built with a subject-orientation and can provide a structure that supports integration over time. A time aspect can be added to the entity key, and, thus, the model easily deals with history. The relational structure further supports having an unbiased source of data for all types of data marts, not just OLAP ones.

In contrast, the dimensional model is built using predefined relationships. These relationships are very powerful for satisfying specific sets of business questions. The predefined relationships may actually hinder answering a different set of questions if these are used to define the primary structure in the data warehouse. The dimensional model can support history through changing dimensions. However,, depending on the types of changes that occur over time, the complexity of the star-schema increases. (In his column, Ralph Kimball devoted several articles to handling history in the star-schema.) Just as a hammer can be used to drive a screw into a piece of wood, a dimensional model can be used to serve some of the needs of a data warehouse. When the hammer is used, the wood often splinters and sometimes cracks. If a dimensional model is used for the data warehouse, the growth potential and applicability of the data warehouse to meet more than the currently known set of business questions is jeopardized.

A relational structure, therefore, is recommended for the data warehouse. This model is based on a third-normal form model that depicts the business relationships. Although the third-normal form model has its place, it is also not the best structure for the data warehouse. Using the third-normal form model for the data warehouse is analogous to selecting any screwdriver for the job. If the screw we have is a No. 4 Phillips head, the best screwdriver is a No. 1 Phillips head. The eight-step process that follows adjusts the third-normal form model to the job at hand.

Step 1 — Select the Data of Interest

The logical model is streamlined to include only the strategic data needed to support the business capabilities for which the data warehouse is being built.

Step 2 — Add Time Aspect to the Key

The incorporation of time into the key of entities facilitates capturing history.

Step 3 — Add Derived Data

Predefined, standard derived data provides consistency across the enterprise for key business data elements, such as net sales amount and profit amount.

Step 4 — Adjust Granularity Level

The level of granularity for the data warehouse supports all of the data marts.

Step 5 — Add Summarized Data

The standard summarized data facilitates delivery of data to the data marts without necessitating redundant calculations.

Step 6 — Merge Tables

With the strategic focus, certain tables are merged to reduce the joins required to deliver or access the data in the data warehouse.

Step 7 — Incorporate Arrays

When appropriate, arrays can further facilitate the delivery of data to the data marts.

Step 8 — Segregate Data Based on Use and Stability

Data in the warehouse is segregated based on its stability (for example, how often it changes) and on the usage commonality. This step helps both in the data acquisition and in the data delivery.

In addition, the data warehouse model supports creating conforming dimensions to further simplify the data delivery process for OLAP marts.

SUMMARY

The key to selecting the appropriate modeling technique is an understanding of the basic function of the data structure. The data warehouse reflects the enterprise business rules, is the source of data for the data marts, and supports direct access only as a secondary objective. The relational model meets these requirements very well. The data mart that is used for OLAP requires easy, intuitive access by business users. The dimensional model is best suited to meet these needs. We have two excellent modeling approaches at our disposal. We need to ensure that we use the right model for each data structure.

CHAPTER 14

The Three Cs of Data Warehousing Philosophy — Commitment, Completeness, and Connectivity

Topics in this chapter:

- Commitment

- Completeness

- Connectivity

- Summary

First, why does the delivery person stop by my house twice in one day to deliver books that were on the same order and were shipped the same day but by different distribution centers? Why does my credit card company keep sending me offers for new credit cards, as if they don't know that I already have an account? Why doesn't my leasing company know that I have two cars leased with them – both in my name – and constantly misapply the payments? Why can't the office-supply company tell me whether it has the pen I want in stock at another store – with the assurance that it is really at the other store, so I won't go on a wild goose chase?

These seemingly disparate problems all have the same root: These companies can't get their acts together because they don't make good use of the billions of bits of information that spew forth from their IT systems and pile up, unanalyzed and unreported, day after day. You wouldn't be reading this chapter if you didn't have, or weren't building, a data warehouse, right? With that in mind, I would like to take this opportunity to give you one piece of advice: make sure that the objective of your data warehouse is to enhance your existing customer relationships and to acquire new customers. That sounds absurdly obvious, doesn't it?

You would be surprised by the reasons my clients give when I ask them why they want to build a data warehouse: more accurate forecasting, more efficient production/operations management, tighter logistics and inventory management, more in-depth financial analysis. The list goes on and on. Did you see the pattern? They all want to know more information about one thing: *demand.* Demand equals *customers.* They all want to know about their customers; they just don't know that it's their customers that they want to know about; at least they don't put it that way. Let's look at obtaining the information we need from the mountains of data we have.

I'll leave the hard-core data warehouse architectural principals to others – for now. What I'm driving at, for the moment, is the philosophy behind the architecture. I'm looking for a form that enables function, not one that follows it. A sound technical architecture may take several physical forms, some of which are better than others. Its logical, or philosophical, form must include three things: commitment, completeness, and connectivity. Without any one of these attributes, the architecture won't do the job. With all three of them, it will be spectacular.

COMMITMENT

The first attribute of a sound technical architecture is commitment – both to the project and to the customer. Commitment to the project means taking the architecture seriously, not just slapping the cheapest, or neatest, tools together and installing them on some really fast machines. Taking the architecture seriously means ensuring that the architecture supports the business processes, which, by the way, should be providing outstanding customer service. This approach also means ensuring that the architecture is scalable for the growth that's sure to follow. Finally, commitment to taking the architecture seriously means ensuring that it is flexible enough to handle integration with new systems that you need to bring online.

Speaking of great customer service, the customer should be the ultimate focus of your technical architecture. The customers' needs – even those they don't yet know they have – should be the answer to every question you have. With an architecture that's committed to customer service, all of the financial, marketing, sales, clickstream, and so on, data that you collect can be transformed into valuable information that will enable you to really understand your customers, not just guess what they might do. However, a technical architecture that has the customer as its philosophical center and that is committed in form and function to providing the basis for superb customer service is only the first step in building an exceptional data warehouse architecture.

COMPLETENESS

The next step in constructing a solid technical architecture is ensuring that the architecture is complete. Of course it's going to be complete; the data warehouse won't work if it's not finished. I didn't say *finished*; I said *complete*. Completeness consists of more than having all the machines in place and all the wires connected. A complete data warehouse technical architecture contains all the tools, technologies, and applications it needs and none that it doesn't. As simple as this sounds, I find that this is almost never the case.

One more piece can always be added to the architecture to make it "perfect," or one extraneous piece can be added, which requires valuable extra maintenance time but provides little, if any, value to the data warehouse.

There are, however, several critical pieces needed to complete the technical architecture and to make it philosophically sound. They are a solid OLAP engine, robust data-mining capabilities, and a strong reporting engine to wrap all the valuable information up in a user-friendly package. You know this. I'm really not saying anything new, but I am looking at it from a holistic perspective. OLAP is not new. Neither is data mining, and reporting is a given, so I won't expend my energy convincing you to get a good reporting tool. However, you would be surprised how many of my clients have told me that they thought that OLAP and data mining were mutually exclusive propositions. When I ask why, the answer is usually, "Don't they sort of do the same thing?"

No! No! No! They are not the same, functionally or theoretically. OLAP answers questions that you know you have, about data that you tell it to search through in preset patterns. For example, an OLAP tool can tell you how many units of your product you sold in a particular month and in a particular region, and it will enable you to compare months and regions. Data mining, on the other hand, enables you to find answers to questions you never thought to ask.

With a data-mining tool, you can run analyses that make it possible to ferret out customer buying patterns and association habits. For example, let's say that you manufacture "smart-cards" that customers use instead of cash at vending machines. You have a thousand vending machines that use your cards and on which you collect card sales data. With data-mining capabilities, you might find out that when people buy grape sodas, they are more likely to buy crackers, and when they buy colas, they are more likely to buy chips. With this information, you can practice targeted marketing and send them e-mail promotions for free chips, or crackers, depending on purchasing frequency. Everybody wins. You get more loyal customers, and they get free stuff.

You can see from the hypothetical examples that you do really need both OLAP and data-mining tools to make a solid data warehouse technical architecture. You can think of the standard business questions, and you need an OLAP tool powerful enough to provide the answers to those questions. You can't, however, think of all the myriad questions and comparisons that data-mining tools enable you to pose. If you have a data-mining tool, it can think of them for you and provide you with critical information that you and your competitors didn't even know existed. If you have the information and your competitors don't, you have a competitive advantage. You still need one more piece for your architecture: connectivity.

CONNECTIVITY

By connectivity, I mean having the right users connected in the right way to the data warehouse. One of the biggest mistakes I see in data warehouse construction is the failure to sufficiently include the potential end users in the process from the inception of the project. Sure, the IT folks know the technical ins and outs of building the data warehouse, but the end users know what information they want to see coming out of it. What's more, they know how they want to see that information. Many of the people who will most frequently use the data warehouse are involved in the sales process – either as direct salespeople or as account managers. Either way, they're on the road more than they're in the office, but they still need to access the data warehouse to do their jobs correctly. Make it easy for them; put the data warehouse on the Web – on an ultrasecure, corporate-wide intranet.

You won't have optimal connectivity until your power users can access your data warehouse via the Web. The safest, best route to a Web warehouse is through an intranet. This is not, or it should not be, a frightening proposition. With an intranet-based technical architecture, your Web data warehouse is a closed, secure system. Only people that are *within* your organization can access the data warehouse Web site. Outsiders – due to a complex system of security, encryption, and firewall technologies – cannot. So, even though you might have offices and people scattered literally throughout the world, they're still a part of your organization; they're not outsiders. The information is accessible within your company only and only those with the capability – which is strictly controlled and guarded – to log onto your network can access your warehouse. You get the best of both worlds: security for your priceless data and flexible, real-time access for your power users. That's true connectivity.

SUMMARY

You need more than the right philosophy to build a successful data warehouse technical architecture. You need competent technical people to select, acquire, assemble, and implement the physical pieces of the data warehouse. However, without following a sound, practical construction philosophy, your data warehouse may become nothing more than a jumble of expensive tools and endless wires.

In my opinion, the three most important pieces of a philosophically sound data warehouse architecture are commitment, both to the project and to the customer; completeness, in terms of which tools and technologies you include; and connectivity, of which the Web is a critical component. With these three pieces in place and with a good technical staff to make the physical data warehouse a reflection of the philosophy underlying it, you will be well on your way to being able to use your data warehouse for the purpose you intended when you built it: to gain an advantage over your competitors and to grow your business.

CHAPTER 15

Data Warehouse Goals and Objectives

Topics in this chapter:

- Part 1: Traditional Decision-Support Systems

- Part 2: Short-Term Objectives

- Part 3: Long-Term Objectives

PART 1: TRADITIONAL DECISION-SUPPORT SYSTEMS

This section examines traditional decision-support systems (DSSs) and the reasons why they have failed to provide complete, correct, and timely information to the organization. Parts 2 and 3 of this chapter describe how short-term and long-term data warehouse objectives address the deficiencies of traditional DSS environments.

Shortly after manual processes were automated, management in all companies and across all industries started to ask for data from their newly automated systems. Their requests were heard and promptly put on the back burner, because the main focus of the time was still on automating processes. However, the hunger for data was so great and the tools for manipulating that data were so limited that it did not take long for data processing to split into two segments: operational systems and decision-support systems.

What happened during the decades that followed is a familiar story to all of us. Decision-support systems could not be built fast enough; a new market for DSS tools was created; businesses started to hire their own programmers; and, more recently, some businesses have begun buying and managing their own technology platforms. Even the term *data processing* underwent some transformations to reflect the shift from automating processes to providing information. Data processing (DP) became information systems (IS), then management information systems (MIS), and now information technology (IT). Newer terms, such as corporate information factory (CIF) and business intelligence (BI), are further indications of this shift.

On the surface, it appears that all information needs are being met and that all participants are satisfied. Or are they? Let's examine what's wrong with today's DSS environment. When the impact of the deficiencies of our traditional DSS environment is really understood, the goals and objectives of a data warehouse environment become very clear.

Departmental Views of Data

The speedy proliferation of traditional DSS environments happened in two ways:

1. *IT technicians dedicated to one or more business units.* This is the most common support model. One or more business units are matched up with an IT unit that is dedicated to solving the decision-support needs for those business units. The IT unit may be a formal unit within the IT organization, or it may just be an informal group of two to three people whose priority it is to serve those business units.

2. *IT technicians hired directly by business units.* With the pervasiveness of client/server, this support model is becoming more and more popular. Analysts and programmers are hired directly by the business units, and, in many cases, they develop and implement their systems on a platform owned and maintained by their business unit.

In both cases, the technicians are exclusively supporting their business units. Independent from each other, they analyze the requirements, define the data, find the most suitable operational system from which to source their DSS, apply the most appropriate spin on the data, and basically cater in all ways to the business unit they are supporting. This results in independently developed, stand-alone systems. It requires little examination to understand that both models result in DSS deficiencies when seen from the organizational point of view.

DSS Deficiencies

As we look at some of the most common DSS deficiencies today, we realize that they are so prevalent and that we have become so accustomed to them that we almost accept them as part of the fabric of system development. Those involved may think of the situation as job security. As long as these deficiencies exist, a need will exist to build another system, write another bridge between two systems, rewrite the system in another language, and buy another tool. However, if we seriously think about the impact of the situation, we realize the tremendous waste of time, money, and resources we have also been accepting for all this time.

Just what is the impact for the most common DSS deficiencies? Here are some examples:

Data Is Not Understood

This applies in a lesser degree to the original users of a DSS who had the system custom built. They probably understood their data, or at least they thought they did. However, new users, either in the same business unit or from other business units, who want to use the data, often do not have the same understanding of it. The data names are often reused and misleading, the data content has changed over the years, and how the data is used in reports may not be self-explanatory. These new users must now take the time to search through documentation, if any exists, or ask other people who have been using the data to explain it to them. Most of the time, these explanations are not documented. If they are documented, the documentation seldom leaves the original business unit, is almost always out of date, and its existence is often not known to everyone else in the organization. The time spent by the new users and the staff assisting them to learn about the data is wasted time.

Users Disagree on Data Definitions

Because traditional DSSs are developed independently of each other by different people, data is interpreted and used per a business unit's view and not an organizational view. Where it is possible and appropriate for different business units to have their own view on the data, the absence of reconciling and capturing these views in metadata leads to arguments among users about what the data means, what its values should be, and how it should be used. Again, a lot of unnecessary time is spent on these arguments.

Reports Are Inconsistent

When users don't even realize that they have a different interpretation and view of data, they may reject reports that show totals or a breakdown that differs from their own. They will further label these other reports, along with the systems that generated them, as bad, wrong, useless, or unreliable. Too often these labels spill over to the innocent staff who developed the other systems and reports. Not only is energy wasted on arguing about who is right and who is wrong, but the bigger damage is done when different groups no longer respect each other, and the work atmosphere is degraded. The impact is lower morale, which usually manifests itself as lower productivity and lower quality.

Users Don't Trust the Reports

This is often the result when the communication between IT and users is limited and when IT develops the system with little or no user involvement, besides gathering the original requirements. Here we have a situation in which the understanding of data is not only different between business units but also between the business unit and its supporting IT staff. To make things worse, not only may IT lack understanding of the business view of the data, but the users often don't understand how the data is being captured and manipulated in the operational system. In this "us versus them" environment, users often try to handle their distrust by creating their own pseudo-IT environment. The impact of this solution surpasses low morale and low productivity and moves into tangible costs for duplicating technology platforms.

Data Is "Dirty"

We hear this complaint frequently from users and IT alike. Yet, when faced with the task of analyzing the operational source data and cleaning it up for the DSS, we also frequently hear excuses, such as

- We are accustomed to the dirty content in this field.

- We are too busy with other things.

- We will just write our queries to eliminate certain records.

- We know how to interpret the bad data values.

- It would take much too long to clean the data.

- It would cost too much to clean the data.

- They will never tighten edit checks on the operational system.

- It's really not that bad once you get accustomed to it.

When confronted with the fact that every other person who does not intimately know the dirty data would need a lot of time to learn how to eliminate certain records or to interpret bad data values, users usually offer two types of responses:

- They'll just have to take the time to learn. That's what we had to do. That's just how it is.

- This is our data. No one else needs to access it. They should call us when they need our data.

Needless to say, the time it takes for many people to relearn the same facts is not being taken into account, and the associated cost is not being considered.

Data Is Not Shared or Is Shared Reluctantly

Because traditional DSS development has been business-unit centric, it is no surprise that data sharing is neither encouraged nor sought. Users who have full control over their systems in terms of data definitions, data acquisition, data cleansing, data transformation, database design, and tools used have no incentive to share what they develop. After all, because the funding for their system came out of their own budget, they have the right to be in full control of what goes into the system, what comes out of it, and who uses it. Certainly no incentives are evident for giving up that control, especially if upper management is still holding them accountable for delivering the system in a very short time frame. Sharing means involving other users. Involving other users means reconciling their views of the data. Considering other views will slow development. Slowed development could result in missed deadlines. Missed deadlines will be remembered at performance appraisal time. An unfavorable performance appraisal will result in a small raise, or no raise at all, and most likely, in no bonus for the year. By keeping things separate, the raises and bonuses are safe, even though different users will spend time and money to reinvent the proverbial wheel.

Data Is Not Integrated

Even if the group were willing to allow other users to share the data, as long as business units build their stand-alone systems based on their individual views, data between systems will not be integrated. Therefore, accessing data across multiple systems often involves writing complicated bridges between systems – a time-consuming and costly solution.

Historical Data Is Not Available

Another paradigm shift exists in DSS: the change from operational decision making to tactical and strategic decision making. The user community of the DSS environment is also changing from business administrators to business analysts and executives from business units such as marketing, legal, finance, and human resources. The new focus on tactical and especially on strategic decision making brings a new requirement to DSS: the capability to compare data between time periods, geographic regions, and other business dimensions. This translates into the need for historical data, which is as easily accessible as current data. Because traditional DSSs do not typically store historical data in the same manner as current data, analysts end up creating new elaborate systems to accomplish their analyses. It can be weeks before an analyst can complete an analysis assignment, because that is how long the process of extracting current data, merging it with historical data, running queries, and analyzing the query results may take each time.

Data Management Solutions

Data warehousing is not the first attempt at tackling these data-management problems. However, if done correctly, it seems to be the most effective so far.

Information Engineering

Data management became a major topic in most organizations in the early 1980s. Information engineering was created as a new IT unit, chartered with developing and applying methods and techniques to manage the organization's data the same way as any other corporate resource is managed.

One of the first methods for managing data was the corporate data dictionary. Information engineers spent years loading a central data dictionary with technical data from their operational systems. They ended up with thousands of data elements, hundreds of files, hundreds and thousands of programs and job control language (JCL) procedures representing all of their systems. It took many more years to analyze and define all the accumulated systems-related data in the data dictionary. This was an honorable first attempt at gathering and maintaining metadata. However, the only visible benefit of this exercise appeared to be a tremendous understanding of corporate data by one or two IT analysts and not much else.

Most companies decided that this was not a cost-effective approach to solving their data management problems, and information engineering is now an affectionate memory of the past.

Data Administration

Because the idea was good, but the concept was unworkable, information engineers quickly reinvented themselves as data administrators. Their mission was the same, but their method was different. A new technique was gaining much popularity during the rise of relational database management systems. This technique was called *entity-relationship modeling* or logical data modeling. Some distinct benefits to the logical data modeling approach over the old data dictionary approach were

Top-down analysis. Logical data modeling was based on user participation. Having users involved in the analysis shortened the process considerably.

Business-centric. The data was being analyzed from a business perspective, not from a technical perspective. The benefit was that users, not just one or two IT analysts, now had a tremendous understanding of corporate data.

Relational design. What really saved data administration from becoming another extinct IT species was the direct applicability of a logical data model to relational database design. With few modifications to a logical data model, we were actually able to implement the model as an application database.

Data administration is still aiming for more than managing the data for isolated stand-alone application databases. The charter of data administration remains the management of corporate data across all business units and all systems in the organization.

Data Warehouse Mission

The mission of a data warehouse is to provide consistent and reconciled business intelligence, which is based on operational data, decision support data, and external data – to all business units in the organization. In order to do that, corporate data must be analyzed, understood, transformed, and delivered. Therefore, the data warehouse administration must coordinate and oversee the development, delivery, management, and maintenance of the entire data warehouse environment.

It has been difficult for IT professionals to keep up with user demands for information. Many DSS projects are running in parallel to satisfy the different business units of an organization. Because these parallel development activities are neither coordinated nor integrated, they are producing stand-alone DSS systems. These stand-alone systems do not provide an acceptable data management solution because of their inherent deficiencies.

PART 2: SHORT-TERM OBJECTIVES

We can now recognize the role a data warehouse plays, or should play, in a data management solution. A data warehouse is not just another DSS database. It is an environment of one or more databases designed to deliver consistent and reconciled business intelligence to all business units in the organization.

To avoid the same calamity that befell information engineering when trying to correct all data management problems in one big-bang approach, you want to separate your data warehouse objectives into two categories: short-term objectives and long-term objectives.

Short-term objectives are those you can realize with every data warehouse iteration. They represent the immediate benefits to the users. Here are some examples of short-term objectives:

Improve Quality of Data

Because a common DSS deficiency is "dirty data," you will have to address the quality of your data during every data warehouse iteration. Data cleansing is a sticky problem in data warehousing. On one hand, a data warehouse is supposed to provide clean, integrated, consistent, and reconciled data from multiple sources. On the other hand, we are faced with a development schedule of six to 12 months. It is almost impossible to achieve both without making some compromises. The difficulty lies in determining what compromises to make. Here are some guidelines for determining your specific goals when cleansing your source data:

Never Try To Cleanse ALL the Data

Everyone would like to have all the data perfectly clean, but nobody is willing to pay for the cleansing or to wait for it to get done. To clean it all would simply take too long. The time and cost involved often exceeds the benefits.

Never Cleanse NOTHING

In other words, always plan to clean some data. After all, one of the reasons for building the data warehouse is to provide cleaner and more reliable data than you have in your existing OLTP or DSS systems.

Determine the Benefits of Having Clean Data

Examine the reasons for building the data warehouse:

- Do you have inconsistent reports?

- What is the cause for these inconsistencies?

- Is the cause dirty data, or is it programming errors?

- What dollars are lost due to dirty data?

- Which data is dirty?

Determine the Cost for Cleansing the Data

Before you make cleansing all the dirty data your goal, you must determine the cleansing cost for each dirty data element. Examine how long it would take to perform the following tasks:

- Analyze the data.

- Determine the correct data values and correction algorithms.

- Write the data cleansing programs.

- Correct the old files and databases (if appropriate).

Compare Cost for Cleansing to Dollars Lost by Leaving Dirty Data

Everything in business must be cost-justified; this applies to data cleansing as well. For each data element, compare the cost for cleansing it to the business loss being incurred by leaving it dirty and decide whether to include that data in your data cleansing goal. If dollars lost exceed the cost of cleansing, put the data on the "to-be-cleansed" list. If the cost for cleansing exceeds the dollars lost, do not put the data on the to-be-cleansed list.

Prioritize the Dirty Data You Considered for Your Data Cleansing Goal

A difficult part of compromising is balancing the time you have for the project with the goals you are trying to achieve. Even though you may have been cautious in selecting dirty data for your cleansing goal, you may still have too much dirty data on your to-be-cleansed list. Prioritize your list.

For Each Prioritized Dirty Data Item, Ask, "Can It Be Cleansed?"

You may have to do some research to find out whether the good data still exists anywhere. Places to search could be other files and databases, old documentation, manual file folders, and even desk drawers. Sometimes the data values are so convoluted that to write the transformation logic, you may have to find some old-timers who still remember what all the data values meant. Then there will be times when, after several days of research, you find out that you couldn't cleanse a data element, even if you wanted to, and you have to remove the item from your cleansing goal.

As you *document your data cleansing goal,* you want to include the following information:

- The degree of current dirtiness (either by percentage or number of records)

- The dollars lost due to its dirtiness

- The cost for cleansing it

- The degree of cleanliness you want to achieve (either by percentage or number of records)

Minimize Inconsistent Reports

Addressing another common complaint about current DSS environments, namely inconsistent reports, will most likely become one of your data warehouse goals. Inconsistent reports are mainly caused by the misuse of data, and the primary reason for the misuse of data is disagreement or misunderstanding of the meaning or the content of data. Correcting this problem is another predicament in data warehousing, because it requires the interested business units to resolve their disagreements or misunderstandings. This type of effort has more than once torpedoed a data warehouse project because it took too long to resolve the disputes. Ignoring the issue is not a solution, either. We suggest the following guidelines:

Identify All the Data in Dispute

Examine carefully how the disagreement or misunderstanding of the data contributes to producing inconsistent report totals.

Determine the Impact of the Inconsistent Report Totals

- How seriously are they compromising business decisions?

- What dollars are lost due to bad decisions?

- Are the differences significant?

- How easy are the reports to reconcile?

Determine the Cost for Resolving Data in Dispute

Estimate how long it would take to

- Get the involved business units to commit to the process of resolving their disputes.

- Analyze the data disputes and model the different user views.

- Separate the consistent views from the inconsistent views.

- Come to an understanding on definitions and content of data for the consistent view.

- Create data for the inconsistent views.

- Come to an understanding on definitions and content of the new data.

Compare Cost for Data Resolution to Dollars Lost by Leaving Data Disputes Unresolved

A cost benefit must be demonstrated before including the resolution of a data dispute in your goal: If dollars lost exceed the cost of resolution, put the data dispute on the to-be-resolved list; if cost for resolution exceeds dollars lost, do not put the data dispute on the to-be-resolved list.

Prioritize the Data Resolutions You Consider Tackling

Anyone who has ever participated in a data resolution session knows how time-consuming it can be. Your project schedule may not enable you to resolve all the data in dispute. Therefore, prioritize your list.

As you did with your data cleansing goal, you want to *document the following information* for your "minimize inconsistent reports" goal:

- The degree of impact on business decisions

- The dollars lost due to data disputes

- The cost for resolving the disputes

- The degree of *resolution* you want to achieve

For example, do all users have to agree or only the two main users? Do the totals have to agree 100 percent or is a 5-percent variance acceptable? If resolution cannot be achieved in X days, can the data be dropped?

Capture and Provide Access to Metadata

Until now, metadata has always been considered the dirty D word: *documentation.* However, metadata is indispensable for data sharing and data navigation.

Data Sharing

Most data is not being shared today for a number of reasons. One reason is not understanding the data, and another reason is not trusting the data content. We already established the fact that in order to correct this problem, users have to discuss their views of the data and discover their commonality and differences. Two major goals of this discussion are commonly agreed-upon data definitions and commonly agreed-upon domains (valid data values). Because these two goals are often misunderstood and declared as unattainable and a waste of time, we must be clear on what we mean by these goals.

The process of achieving commonly agreed-upon definitions and domains does not mean that hundreds of users are arguing ad infinitum about who is right and who is wrong, and the desired result is not a declaration of victory by the most powerful user having forced his or her opinion on the other users. The process involves a small group of people, usually five or six, consisting of a facilitator, the data owner, and one authoritative representative from each business unit, which is using the data for making significant business decisions. When strong disagreements surface about the meaning or content of the data, it is an indication of high probability that all of the disagreeing parties are right and that more than one data element exists. This probability is explored within a predefined short time frame, usually no more than a few days; and a new data element is created, named and defined by the group. If the exploration does not yield a new data element, the data owner makes the final decision on definition and content. Assuming there is no turf war between the disagreeing parties, the definition and content will, to some extent, include any reasonable variations perceived by the other involved business units. The now agreed-upon definitions and content for the original and for the new data elements are documented in a metadata repository and made available to all other users in the organization.

Data Navigation

We like to think of metadata as the nice N word: *navigation.* After the source data has been cleansed, transformed, aggregated, summarized, and dissected in numerous other ways, the users will never find it again in the data warehouse without the help of metadata. Capturing the metadata (for example, the data definitions, the domains, the algorithms for transforming the source data, the columns and tables in which the resulting data resides, and all the other technical components) is only half of the solution. The other half is making metadata easily accessible and useful to the users.

Provide Capability for Data Sharing

If data sharing is one of your data warehouse goals, you also must include some data cleansing, data dispute resolution, and metadata access components as means for achieving this goal. These components are prerequisites to data sharing. Two other vital components are database design and database access.

Database Design

After the requirements have been analyzed, the requested data has been logically modeled, and the related metadata has been captured in the repository, the next step is database design. Designing a stand-alone database for one business unit is different than designing a shared database for multiple business units. It isn't just a matter of granting access to more users, but a matter of designing a database based on 1) the lowest level of detail and granularity necessary to satisfy all the different data needs, and 2) the type of access required by the different business units.

You can make many design choices depending on how you mix the requirements. When you define data sharing as a goal, you must be specific about the following:

- Technical literacy level of users

- Business knowledge

- The level of detail data required by all users

- The types of summarization and aggregation requested

- The types of queries each user will write

- The periodicity needed (for example, daily, weekly, or monthly snapshots)

Database Access

As with metadata, getting the data into the database is only half the battle. Providing easy access to it is the other half. Not all users are created equal. Some power users may even qualify as programmers, and some technophobes need pull-down menus and radio buttons to navigate. All the competency levels in between also exist. You need to accommodate a wide spectrum of users with a variety of query and reporting tools.

When you document your data sharing goal, describe the users in terms of the following:

- Overall technical literacy level

- The types of queries they are capable of writing

- Whether they will need to manipulate query results

- What summary views they need

- What ad hoc versus report-writing capabilities they need

- How often they will access the system

- Whether they have prior experience with a query or reporting tool

- Their proficiency level with a query or reporting tool

- Their ability and the speed with which they learn new tools

This information will be most valuable for tool evaluation and selection, as well as for training.

Integrate Data from Multiple Sources

Integrating data is another primary goal for all data warehouses, because it is a primary deficiency in current DSSs. A frequent lament is, "It takes me days to merge data manually from four different systems because there is no common key among the files." Stand-alone systems, which have the same data identified by different keys, is only one of many reasons why data integration does not exist in most companies. Some other reasons are that the data content in one file is at a different level of granularity than the data content in another file or that the same data in one file is updated at a different time period than that in another file. In a shared data environment, the requirements from different business units regularly include data relationships, which do not exist in current systems. This often means that the necessary foreign key to implement the requested relationship does not exist in the source files.

Before you define your data integration goal, review your current DSS deficiencies and analyze the source systems you have identified as possible feeds to your data warehouse. Document the following:

- Whether the keys for the same data have the same data type, length, and domain

- Whether the same data is identified by the same key value

- Whether new data relationships can be implemented

- The granularity of the data content

- Periodicity of data updates

Merge Historical Data with Current Data

A typical data warehouse objective is to store history. This objective comes with its own challenges. Historical data is seldom kept on the operational systems, and, even if it is kept, rarely will you find three or five years of history in one file. First, historical data is not as useful to the daily operational processing of a business function as it is to decision support. Second, operational files change over time, and reloading historical data to match the new records' layouts would not be cost-justified. Third, operational history is a point-in-time transaction history, not a periodic snapshot in time. *Point-in-time* transaction history means that a record is written to the file each time a transaction (change) occurs. *Periodic snapshot* means that a record is written to the file once for each period (daily, monthly, and so on) regardless of how many transactions occurred within that period.

Having said all that, you must define the following detail for your goal to merge historical data with current data:

- The number of years for which you want to keep history

- Whether history will be collected from this point (initial loading) forward or whether you will load *X* number of past years

- Whether the history files have the same record layout

- Whether the format of the data has changed over time

- Whether the meaning of the domain (valid values) has changed over time

- Whether the organizational hierarchy has changed over time

- How much history is actually available on disk or on tape

Keep Objectives Realistic

It is important to have clear objectives for a data warehouse, and it is equally important for these objectives to be realistic and cost effective. In addition, objectives should be prioritized, because your project schedule may not enable you to accomplish them all.

PART 3: LONG-TERM OBJECTIVES

The long-term data warehouse objectives resemble many of the original data management objectives from the early 1980s. If you focus on your short-term objectives during your data warehouse iterations, your long-term objectives will almost assuredly be realized.

Here are some examples of long-term data warehouse objectives:

Reconcile Different Views of the Same Data

If your short-term goals include minimizing inconsistent reports and providing the capability for data sharing, you are already addressing this reconciliation effort to some degree. You may have to make suboptimal choices from an organizational perspective in order to complete your project within the scheduled time frame and within your budget, both of which were based on your cost-benefit analysis. This means that you may not achieve comprehensive reconciliation of different views one project at a time. However, as your data warehouse grows with each iteration, increasingly different views of the same data will be addressed and resolved.

If your organization is committed to achieving the highest level of maturity in terms of data management, your data administration unit is most likely chartered with the task of reconciling the remaining different views of the same data. However, this activity is outside the scope of a data warehouse project. As project manager, your involvement in this activity ends with your communication to data administration about the following:

- All the different views of the same data you have discovered

- The views you are planning to reconcile in your project

- The views you are not addressing in your project

Provide a Consolidated Picture of Enterprise Data

One of your project deliverables will be a logical data model of the data within the scope of your project. As you add new data and new requirements to the data warehouse in future iterations, you will expand the logical data model. Over time, this model will grow into a consolidated picture of enterprise data within the scope of the data warehouse.

An organization trying to achieve the highest level of maturity in terms of data management will have chartered the data administration unit with the task of completing that picture. We have to remember that even a fully completed and populated data warehouse will not have all of the operational data used by the organization. To complete the picture, data administration will create or obtain the logical data models from all other systems and merge them into a high-level enterprise model. Clearly, this activity is outside the scope of a data warehouse project. As project manager, your only involvement is to share your logical data model with data administration.

Create a Virtual "One-Stop Shopping" Data Environment

In order to achieve the following long-term objectives, two types of architectures must be very well designed and redesigned with every data warehouse iteration: the technical architecture and the data architecture.

- All the data in the data warehouse environment is accessible through one common interface or point of entry.

- A suite of standard query and reporting tools is available and easy to use.

- The physical location of the data is transparent to the users.

- The data is integrated, clean, and consistent – or at least reconcilable.

- The query results are consistent – or at least reconcilable.

Technical Architecture

Technical architecture has many parts, but in this context, we will address only three: the application layer, data access tools, and database structure.

Application Layer

This is a storefront, a point of entry. It can be a home-grown application written in a conventional programming language with a desktop icon to launch it, a purchased software package, or a Web application. All data warehouse components are hooked into it and are launched from it. This type of an application layer could even be expanded to provide user access to systems other than the data warehouse.

Data Access Tools

The second tier in this architecture is a suite of data access tools. This tier should include a query library of prewritten queries using these tools or written in native SQL.

Database Structure

Also part of the second tier are the actual database files. Their physical placements on one server or another, in one physical location or another, will be completely transparent to the users. All the communication and synchronization between the physical files are handled by the database management system (DBMS) or middleware.

If these three components are designed properly, all the data should be accessible through one common interface, a suite of standard tools will be easy to use, and the physical placement of data will be transparent to the users.

Data Architecture

Data architecture has two parts: the logical data architecture, which is a paper model; and the physical data architecture, which is comprised of the actual physical databases.

Logical Data Architecture

Data issues are addressed at the logical level with a logical data model and metadata. The logical data model will help attain data integration and data sharing; the metadata will help achieve sanctioned definitions and domains.

Physical Data Architecture

Access and performance issues are addressed at the physical level with the appropriate database designs. Because one logical data model can, and often will, be implemented as two or three differently designed databases, it is important to capture the mapping between the logical and physical data architectures as metadata.

If these two data architectures are well designed, all the data will be integrated, clean, and consistent, and all query results will be consistent – or at least reconcilable.

All of your short-term and long-term data warehouse objectives are of no value if they don't match the strategic goals and objectives of your organization. Let's say that one of your data warehouse objectives is to integrate data in order to get a better understanding of your customers' buying habits so that you can sell them more profitable products. On the other hand, your company's strategic goal is to reduce production costs. With this kind of mismatch between company goals and data warehouse objectives, you will not get upper management support for your data warehouse.

The strategic goals of an organization usually revolve around eliminating some business pain. That business pain could be loss of revenue, inability to keep up with competitors, high production costs, or the inability to expand the market share. The development of a data warehouse must support the strategic goals of your organization.

As you are identifying the goals and objectives for your data warehouse project, keep asking yourself: What is the business pain? Do my objectives match the business pain? Will the data warehouse contribute to the relief of that pain? To build a costly data warehouse just to have one or because it is a trendy thing to do is an unwise business decision.

Although some of the long-term data warehouse objectives can be seen as a natural outcome of implementing the short-term data warehouse objectives, other long-term objectives, especially those that span across the organization, will require significant participation from data administration. But, most importantly, *all* data warehouse objectives, whether short term or long term, must be aligned with the strategic goals of the organization.

CHAPTER 16

The Next Big Thing for Data Warehouses

Topics in this chapter:

- Forecasting versus Budgeting

- The Changing Role of the Data Warehouse

- The Changing Role of Business Intelligence

- Extending the Boundaries

- The Language of Collaboration

One report is common to virtually every enterprise and is eagerly antici-pated at the end of each month. The report not only reaches the highest levels of the enterprise's management but is often the catalyst for business analysis across departments and throughout the enterprise. The report compares actual performance with the planned performance.

Surprisingly, this report is rarely produced from data stored in the data warehouse. The simple reason is that plan data is seldom included in the data warehouse design.

Why do so few data warehouses contain plan data? Or, more importantly, what role *should* the data warehouse play in the business planning process? *Should* business intelligence applications play a role in the business planning process?

Even current, state-of- the-art business intelligence applications, backed by the best data warehouses, lack important capabilities that are required to support the process of decision making and decision implementation. Certainly, business intelligence systems excel in providing users with valuable insight into what happened. Decision-making value comes from the analysis and simulation steps required to evaluate alternative decisions and to influence what will happen.

A more significant shortcoming of the data warehouse and business intelligence applications is the support provided for *decision implementation.* Decision implementation is a collaborative process involving many people throughout the enterprise and who are often external to the enterprise. Decisions must be approved and communicated to everyone who must take action. In order to support decision implementation, information must flow back into the operational and financial systems that serve as the sources for the data warehouse.

The linking of decision making and decision implementation establishes a closed-loop business management process. The data warehouse's role becomes more vital to the entire process of business management.

The data warehouse is the single, integrated source of data for decision making. See Figure 16.1. The next big thing for data warehouses is the support of the collaborative planning functions essential for decision implementation. The most actionable data, uniting marketing, finance, and operations, is the demand forecast. A significant change in the forecast affects virtually every business management discipline. The forecast is the language of interdepartmental collaboration.

The Internet enables suppliers and vendors to collaborate more easily to achieve a new level of supply-chain efficiency. An essential goal in this

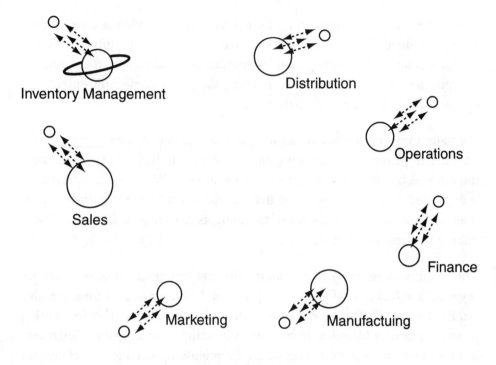

FIGURE 16.1.

Today, most organizations maintain multiple, unrelated forecasting systems.

business-to-business collaboration is reaching consensus on the forecast. Again, the demand forecast represents the common language for collaboration among members of the trading community.

FORECASTING VERSUS BUDGETING

Business planning has two interrelated activities: Forecasting is the projection of future demand, and budgeting is the consolidation of expense line items. The forecast is used to estimate revenue for budgeting; however, forecasts are also critical to operational systems used by purchasing, manufacturing, inventory management, logistics, and so on.

Budgeting is an iterative process of entering and consolidating data, usually with predefined financial goals. Executive management establishes the annual revenue and income goals. Departmental managers work through revenue and expense estimates, making the appropriate adjustments, to operate within the defined goals.

Virtually every organization has a highly disciplined process – not always an efficient process – for creating and consolidating budgets. Consolidated, the budgets become the enterprise's financial plan. When the financial plan is approved, it becomes the yardstick against which the enterprise's performance is measured. The details of the budgets are some of the most closely guarded secrets of the enterprise.

The demand forecast is a key input into the budgeting process, because revenue is a function of unit sales, price, and discounts. Forecasts are also required to support a wide range of business operations. The forecasting process considers historical trends, market factors, and subjective judgment in projecting future sales. Forecasts are frequently updated to reflect current business conditions, because operational systems must react quickly to market conditions. Although an organization generally consolidates a single enterprise budget (the financial plan), forecasting systems proliferate.

Another key difference between forecasting and budgeting is the frequency of updates. Ideally, the budget is created annually and remains unchanged throughout the year. Practically, the budget is updated to reflect major changes in the business; however, often the annual financial goals are not altered. Forecasts are created regularly throughout the year. One area for improving forecast accuracy is increasing the frequency of forecasting. For example, a major supplier of perishable ingredients for the manufacture of food products forecasts the demand from 2,000 customers twice monthly. This company is considering increasing the frequency to weekly.

As a decision support tool, the budget establishes a framework for expense controls and identifies areas for expense reduction if revenue goals are not met. Organizations have financial analysts that continuously analyze performance and identify opportunities for improving financial performance.

The value of the forecast is in supporting the operational decisions that an organization makes each day. In other words, forecasting is less a reflection of what management *wants to happen* and more of a reflection of what is *most likely to happen.* A critical requirement in forecasting is to provide an early warning of significant revenue variances.

Forecast error is computed by comparing actual performance to a prior forecast. Forecast error is generally acceptable at a summary level, the level at which it is translated into the enterprise's annual revenue projection. Forecast error increases at a lower level, the level at which operational decisions are made. In most companies, forecast error is unacceptably high at this level. The economic impact is felt both in terms of high operational costs and poor customer service.

Why make this distinction? Many organizations are satisfied with their budgeting systems. However, increasingly, the brutally efficient nature of business today requires companies to dramatically improve operational efficiencies. Not only can the data warehouse play a key role, but the Internet increasingly offers the opportunity for interenterprise collaboration in consensus forecast development. The economic impact is large and measurable.

THE CHANGING ROLE OF THE DATA WAREHOUSE

Arguably, the best source of data to support the demand forecasting process is the data warehouse. The data has been subjected to considerable cleansing in order to accurately reflect consistent historical trends. The data is also extracted from a number of sources, providing a rich analytic base for cause-and-effect analysis. The one thing that statisticians agree on is that the weakness of a sophisticated statistical forecasting model is the lack of reliable data.

For many reasons, the budgeting process may remain external to the warehouse. However, the data warehouse and business intelligence should play a role in sales forecasting. Much as the data warehouse provides a single source of management information about the past, it should also provide information about the future.

The data warehouse was created because of the need to separate user-accessible data for business intelligence from the transactional data required by operational systems. Bill Inmon's original definition of a data warehouse was "… a subject-oriented, integrated, time-variant, nonvolatile collection of data in support of managements' decision-making process" (*Building the Data Warehouse*, John Wiley & Sons, 1992–1993). This definition does not specify that the data warehouse is intended to be read only. Inmon chose his words carefully in using the term *nonvolatile*.

Maintaining a forecast in the data warehouse requires periodically updating the tables or, more appropriately, creating tables that reflect a new forecast version. This translates into providing users with limited read/write access to portions of the data warehouse. In order to enforce the nonvolatile requirement, a method of committing each forecast version is required. As each new forecast version is developed, users are granted limited read/write access. After the forecast is committed, it becomes a nonvolatile version that can be used for reporting and analysis.

THE CHANGING ROLE OF BUSINESS INTELLIGENCE

The goal of business intelligence is to answer business managers' endless streams of questions. The answer to one question almost always generates a new question. Each new question requires more comprehensive analysis. Many business-intelligence tools focus on query generation and reporting data with light analysis capabilities.

As organizations become successful in answering managements' easy questions, they must anticipate more difficult questions. The hardest questions to answer are: Why? What if? Answering these questions is essential but requires sophisticated data modeling and projection tools. This is not to suggest that all business managers require statistical analysis and data-mining tools. A few might, but most want the application of the technology delivered as a solution to their business issues. In other words, as business intelligence matures, it will take on the form of solutions to business issues.

The application of the technology of business intelligence is already addressing issues associated with customer relationship management (CRM). Advanced business intelligence technology is required for the analysis of customer behavior and market segmentation. A second application direction for business intelligence is demand forecasting. Improving operational efficiencies within the enterprise and across the extended supply chain is one of the most important business issues confronting organizations today. As noted earlier, markets are becoming brutally efficient, and organizations must become equally efficient to survive.

Demand forecasting addresses the what-if questions that are essential to business planning. Forecasting requires statistical modeling and projection technology as well as the capability to override the statistical forecast. A wide range of statistical techniques from simple projection of trends to advanced causal modeling is available. The selection of the appropriate statistical model requires careful analysis. Statistical forecasting is useful but must be supplemented with a management review process. The reason is that most statistical modeling looks at history as the primary indicator of what will happen. An organization's marketing and sales managers are working hard to substantially improve upon history.

Business managers must be able to simulate various scenarios in projecting the impact of their decisions and to change the forecast when appropriate. Ideally, agent processes that issue alerts to managers only when there is a need for review should drive the forecast review process.

Forecasting is a business process that should involve many people throughout the enterprise and between businesses. Business intelligence needs to support each step of the process: analysis, modeling, review, and the publication of forecast versions to operational systems. To do this, business intelligence must encourage greater user collaboration associated with the business management processes.

EXTENDING THE BOUNDARIES

Internet technologies are moving rapidly from the business-to-consumer (B2C) form of e-commerce to the creation of applications for business-to-business (B2B) e-commerce. Ultimately, the challenge will be to go beyond brokering collaboration between two companies, extending to everyone within a trading community (BXB). Information is at the core of interenterprise collaboration. More specifically, the demand forecast is the language of collaboration between businesses. The one thing on which both supplier and vendor must reach consensus is the forecast; otherwise, neither can achieve optimal levels of efficiency.

The data warehouse and business intelligence must operate outside of the boundaries of the single enterprise. Business intelligence must be Web optimized for ease of delivery, and issues such as security and scalability become extremely important.

Certainly, extending the boundaries of data warehousing and business intelligence has the potential to change the rules for how a company operates and competes in what can only be termed a fast world. The non-trivial challenge is positioning the data warehouse and business intelligence in the center of what a business requires to compete in the fast world (Figure 16.2).

THE LANGUAGE OF COLLABORATION

The next big thing is the integration of the data warehouse and business intelligence with business management processes. Internet technologies promote a new level of global interdepartmental and, increasingly, interenterprise collaboration. The language of business collaboration – for both interdepartmental and interenterprise – is demand forecasting.

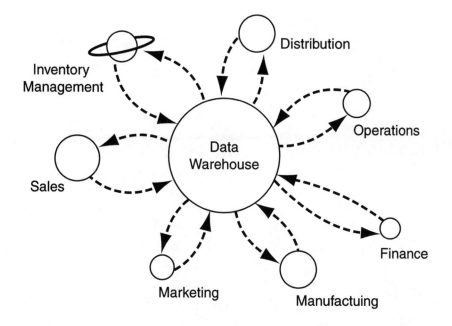

FIGURE 16.2.

The next big thing is for the data warehouse to become a read/write tool at the center of your universe. All entries work from this single source, creating a collaborative demand forecast.

CHAPTER 17

Various Tools in Data Warehousing

Data Warehousing Solutions— What Is Hot?

Solution Area	Product	Vendor
Report and query	Impromptu	Cognos
	BrioQuery	Brio Technology
	Business Objects	Business Objects, Inc.
	Crystel Reports	Seagate Software
OLAP / MD analysis	DSS Agent/Server	Microstrategy
	DecisionSuite	Information Advantage
	EssBase	Hyperion Solutions
	Express Server	Oracle Corp.
	PowerPlay	Cognos Corporation
	Brio Enterprise	Brio Technology
	Business Objects	Business Objects

(continued)

Solution Area	Product	Vendor
Data mining	Discovery Server	Pilot Software
	Intelligent Minor	IBM
	Darwin	Thinking Machines
Data Modeling	ER/Win	Platinum
Data extraction,	DataPropagator	IBM
transformation,	InfoPump	Platinum Technology
and load	Integrity Data Re-Eng.	Vality Technology
	Warehouse Manager	Prism Solutions
	PowerMart, PowerCenter.	Informatica
Databases for data	DB2	IBM
warehousing	Oracle Server	Oracle
	MS SQL Server	Microsoft
	RedBrick Warehouse	Red Brick Corp.
	SAS System	SAS Institute
	Teradata DBS	NCR
Information	DataGuide	IBM
catalogue	HP Intelligent Warehouse: Guide	Hewlett-Packard
	Directory Manager	Prism Solutions

CHAPTER 18

Building a Data Warehouse — A Few Salient Points

Topics in this chapter:

IMPLEMENTATION VIEW OF DATA WAREHOUSE APPLICATIONS

When building a data warehouse application from the bottom up rather than the top down, it is important to start off in a simple manner, to stay focused on the goal, and to prove every step that you take throughout the development process. Every step should be proven, and you should have methodologies of work to show for proof.

Use an incremental, multiphase approach that entails the following:

- Defining the business drivers, compiling a complete requirements list or a requirements study document, and creating a specification document that reflects the needs of the end user

- Defining the data warehouse architecture, the various components, and the tools being used therein

- Developing a pilot of the project for the initial architecture of the data mart – also called a Rapid Application Development (RAD)

- Expanding the pilot to view the whole architecture globally and plugging in additional data marts

- Evolving the enterprise data warehouse architecture

- Administering and maintaining the data warehouse

DEVELOPMENTAL STRATEGY

There are two development strategies in use today: the Bottom-Up and Top-Down.

Bottom-Up Strategy

This strategy also is called the *Incremental Development Strategy,* and it is the most recommended of all approaches today. This strategy involves the following:

- Building the pilot project as a single architecture data mart

- Building additional data marts with one business area at one time

- Building a central data warehouse to store the detailed data

- Reducing the costs and risks and supporting the early deployment of the first data mart, resulting in a rapid ROI

Three phases are involved here.

Phase 1

Design the long-term enterprise data warehouse framework and build the initial architectured data mart, using the exact subset of enterprise architecture, including a scalable extraction/transformation/loading tool and a central metadata repository.

Phase 2

Build architectured data marts for additional subject areas. Ensure that each integrates with the central metadata.

Phase 3

Expand the data warehouse to enterprise model. Implement the central data warehouse to store the detailed data.

Top-Down Strategy

This strategy is based on a subset of the parent data warehouse and involves the following:

- Building the central data warehouse first

- Then building the initial data mart using a subset of data in the central data warehouse

- Building additional data marts using subsets of data in the central data warehouse

- Implementing data marts as exact subsets of the central data warehouse

This strategy is not recommended because of the time involved and the cost of building the central data warehouse, and there is a lengthy delay for deployment of the first data mart.

Phase 1

Design the long-term enterprise for the data warehouse framework. Build the central data warehouse, including a superset of all dimensions, attributes, aggregations, and so on, for all areas.

Phase 2

Build the initial data mart using a subset of data.

Phase 3

Build additional data marts, all of which inherit dimensions, attributes, metadata, and so on from the parent.

Recommended Implementation Techniques

To define the structure of the enterprise data warehouse and architected data marts

- Build the data warehouse incrementally, one business area at a time.

- Buy only components that integrate with the central metadata repository.

- Ensure that the data warehouses are populated with bad data.

- Support a mix of RDBMS, Multidimensional Databases (MDBs), and hybrid target databases.

- Provide easy-to-use tools that do not require assistance; this means using templates.

- Ensure that tools provide the same functions on a LAN and Web.

- Use the Web to reduce the cost of communications and HW and SW support.

- Support mobile users with offline query, reporting, and OLAP functions.

- Ensure that the system is scalable to increases in users and database size.

- Provide for powerful security and warehouse administration functions.

BUILDING THE DATA WAREHOUSE

1. Identify the business drivers, sponsorship, risks, and ROI.

2. Survey user needs; identify the desired functionality.

3. Design long-term, enterprise data warehousing architecture on paper.

4. Define functional requirements for initial subject area.

5. Research and select DSS components and tools. Conduct *Proof-of-Concept* test.

6. Design target database.

7. Build data mapping, extract, and transformation rules.

8. Build aggregation, summarization, partition, and distribution.

9. Build the initial architected data mart, using the exact subset of enterprise data warehousing architecture.

10. Build additional architected data marts.

11. Expand to enterprise architecture, including central data warehouse and optional Operational Data Store (ODS).

12. Maintain and administer the data warehouse.

Drivers, Sponsorship, Risks, and ROI

1. Build the data warehouse solely to solve a business problem.

2. The business problem must be specific and painful:

   ▼ Company is losing its competitive advantage.

   ▼ Customers are moving to competitors.

   ▼ Management has little insight and control over costs.

▼ Promotions are failing for unknown reasons.

▼ Turnover of goods is low, and cost of inventory is high.

▼ Business managers have an inadequate understanding of customer needs.

Survey User Needs

1. What is the mission and responsibility of the business unit?

2. What are the challenges facing the business unit?

3. What OLTP and DSS business processes are in use?

4. Are there specific problems with the current DSS systems?

5. What type of users must be supported – for example, query/reporting, power users, financial analysts, executives?

6. What are the business rules and semantics for the business area? Are they consistent across the unit?

7. What are the defined common business rules, semantics, definitions, facts, metrics, dimensions, and attributes?

Questions To Be Answered in User Needs Workshop

1. What are the subject area and the business drivers for the data mart?

2. Source systems – where does the detail data come from?

3. Atomic level of fact detail – what level of detail?

4. Length of fact detail history – how long to retain detail?

5. Required business dimensions and attributes – what are the column headers on reports?

6. Multidimensional aggregation requirements – what combination of views or slices are required by users?

7. History tables – does the user need to tack changes in dimensions and attributes?

Design Enterprise DW Architecture on Paper

1. Define the DW architecture in a one- to two-day JAD workshop.

2. Design all components of the data warehousing application:

   ▼ Data extraction/transformation/load tool (ETL)

   ▼ Central metadata repository

   ▼ Data warehouse for detail data

   ▼ Central data warehouse for detail data

   ▼ Multiple architected data marts

   ▼ Data access and analysis tools for end users

   ▼ Integration of DW components with central metadata

Requirements for the Initial Subject Area

1. Define the development methodology – bottom-up or top-down approach. The bottom-up approach is recommended.

2. Identify the initial business area to be supported.

3. Conduct a requirements planning JAD workshop to define the functional requirements, tasks, phases, project plan, pilot project, schedules, budget, resources, skill sets, and so on. The objective of the JAD workshop is to gain consensus on all issues related to building the initial data mart.

4. Following the resolution of all issues, immediately begin the implementation of the initial architected data mart.

Select Data Warehousing Components and Tools

The components to be selected include the following:

- Extraction/transformation/load tool (family of tools including data modeling tool, extraction tool, metadata repository, and DW administration tools)

- Metadata exchange architecture (API used to integrate all components of DW with central metadata)

- Target databases (relational, multidimensional, and hybrid)

- Data access and analysis tools for end users

- Database servers, operating systems, and networks

Proof-of-Concept Document

Prepare a document with the following sections:

- Specification of source tables/fields to be extracted

- Specification of target data model

- Specification of source-to-target mappings

- Specification of at least three aggregates

- Specification of session and management commands

- Specification of at least three end-user reports

The document will take about two weeks to prepare. Present document to vendors of ETL and end-user tools.

Design the Target Database

Design a logical data model for the target database:

- Use data facts, dimensions, and attributes as defined earlier.

- Design the logical data model for all target data entities.

- Import data models from external sources or design them with the data modeling component of an ETL tool.

Design the physical model by denormalizing the logical model:

- Star-schema for relational target database (avoid using a snowflake schema to simplify end-user browsing)

- Multidimensional array for multidimensional database

Build Data Mapping, Extract, and Transformation Rules

1. Use current reports to define the initial required entities.

2. Identify source data for each required entity and map source data to target dimensions and attributes.

3. Define data cleansing rules.

4. Define data extract and conversion rules:

   ▼ Name changes, key changes

   ▼ Physical attribute changes

   ▼ Filters, defaults, and reference table conversions

   ▼ Multiple-choice selections (many to one)

Data Cleansing Process

1. Measure data quality: Identify data with inconsistent, missing, incomplete, duplicative, or incorrect values.

2. Standardize, integrate, and identify duplicative name and address data and household customer data.

3. Identify multiple occurrences of the same individual or business in various source systems.

4. Ensure that the Customer Dimension is clean.

5. Clean data at the source or as part of the ETL process.

6. Load only clean data into the data warehouse.

7. Identify and correct the cause of data defects.

Build Aggregations, Summarization, and Partitions

1. Identify commonly used multidimensional aggregations (precomputed data cubes) – for example, sales by region by customer by product by month.

2. Precompute aggregation for fast query response.

3. Implement aggregations and summarizations using the point-and-click interface of an extract/transformation/load tool.

4. Define the required level of granularity of data.

5. Partition the data by month, region, and so on, if necessary.

Steps 4 and 5 may require 70 percent of your effort to build a data mart.

Build Initial Architected Data Mart

1. Implement the initial architected data mart using an exact subset of the enterprise. For the DSS architecture,

   ▼ Include an extraction/transformation/load tool and data modeling tool.

   ▼ Keep the initial transformations simple.

   ▼ Build the initial data mart for one business area.

   ▼ Select a simple desktop OLAP tool for end users.

2. Deliver the first data mart three months after the initiation of the project.

3. Ensure a rapid return on investment from the initial data mart.

Build Additional Architected Data Marts

1. Identify areas of business that require data marts.

2. Survey user needs (additional dimensions and attributes).

3. Define the architecture of additional data marts.

4. Define functional requirements of additional data marts.

5. Design the target database for additional data marts.

6. Build data mapping, extract, and transformation rules.

7. Build aggregations, summarizations, and partitions.

8. Ensure that all components integrate with the central metadata.

9. Deliver additional data marts every three months.

Expand to Enterprise Data Warehouse

1. A large, central data warehouse drives multiple architected data marts.

2. The central data warehouse stores detail data (individual transactions).

3. Expansion is required when end users need access to detail data.

4. The enterprise data warehouse supports organization-wide consolidated analysis, reporting, and queries.

5. The enterprise data warehouse is a complex environment with high developmental cost and risk.

Maintain and Administer Data Warehouse

1. Data warehouse system administration is an ongoing process.

2. Use DW administration tools to set up users, authorize access, and set security.

3. Monitor access and usage patterns.

4. Block long queries and reschedule them to run in batches.

5. Monitor business issues facing end users so that queries can be predicted and planned for in advance.

6. Use DW data administration tools to restructure physical database structures to improve performance.

Summary

1. Identify business drivers of data warehouse.

2. Use DW to solve painful, strategic business problems.

3. DW must be business-driven, not technology-driven.

4. Do not load dirty, inconsistent data into the DW.

5. Do not build stovepipe data marts. Do not buy any tool that cannot be integrated with central metadata.

6. Build data warehouses from bottom up, not top down.

7. Build the pilot project as an exact subset of enterprise DW architecture to minimize cost and risk and to ensure rapid ROI.

8. Ensure that DW is scalable and meets performance goals.

RELATIONAL, MULTIDIMENSIONAL, AND HYBRID TARGET DATABASES

1. Conventional RDBMSs may be used to support most data warehousing requirements and can handle very large target databases.

The central data warehouse is almost always a conventional RDBMS.

2. Multidimensional databases (MDBs) provide faster response to analytical queries and OLAP computations than RDBMSs, but they have size limitations.

3. The hybrid approach combines relational and multidimensional technologies within a single database. Relational technology is used to store large database segments; multidimensional technology is used to provide fast response to analytical queries

4. Hybrid databases are likely to be widely used in the future.

Strengths of Conventional RDBMSs for the Data Warehouse

1. No size limit is placed on the target database (scalability to multiterabytes).

2. The standard corporate database can be used for building the data warehouse. No need for proprietary databases and tools.

3. RDBMSs have been enhanced to support the specialized requirements of data warehouses. Techniques include parallel query execution, parallel data management, cost-based query optimization, bitmapped indexes, SQL extensions for OLAP, and so on.

4. Ad hoc queries can be directed to any data in the data warehouse.

5. Detailed data and aggregated data can be stored in the same database.

6. High-speed query processing is possible using symmetric multiprocessors (SMP), massively parallel processors (MPP), and clustered multiprocessors (NUMA).

Limitations of Conventional RDBMSs for the Target Database

1. Originally optimized to support on-line transaction processing

2. Typically slower than MDB databases for OLAP calculations and ad hoc analysis of data in multiple dimensions

3. SQL not designed for calculations and OLAP functions

    ▼ SQL cannot calculate variances between rows or perform multilevel hierarchical roll-ups.

    ▼ SQL cannot perform ratios, comparisons, what-if functions, rankings, consolidations, or cross-dimensional calculations.

4. Sophisticated techniques are required to generate SQL, process queries, and perform aggregation and calculations, in order to overcome SQL limitations (for example, multipass SQL, temp tables).

5. *Read-only.* Does not support read-write operations for budgeting

What To Look for in a Relational Target Database

1. Scalability to support very large databases (terabytes) and large numbers of concurrent end users performing complex analyses

2. Adequate performance for ad hoc queries to any date in the database

3. Support for advanced parallel processing techniques

4. Evolution to a hybrid database architecture

5. Integration with data ETL tools

6. Integration with multidimensional databases

7. Support for star join and multidimensional extensions to SQL to support OLAP calculations, variances, moving averages, and so on.

8. Portability, security, data integrity, backup/restore

Representative Relational Database Management Systems

1. Oracle 7.3, 8.x Parallel Server supports shared databases, not MPP.

2. IBM DB2 Universal Database for MVS, AS/400, AIX, NT, OS/2, HP/UX, Sun Solaris, and SCO. DB2 Parallel Edition runs on IBM SP2.

3. Informix Extended Parallel Server (XPS), Internet Foundation 2000 (was Universal Serbver), and Decision Frontier Suite

4. Microsoft SQL Server 6.5, 7.0 (SQL Server OLAP Services)

5. Terradata from NCR runs on NCR WorldMark Server (MPP).

6. Sybase Adaptive Server Enterprise, Adaptive Server IQ, and Sybase Warehouse Studio

7. Red Brick Decision Server from Informix Software (data marts)

When Should You Use a Relational Target Database?

1. Requirement to support *ad hoc* (unpredictable) queries to any portion of a very large target database

2. Constraint to use only the standard corporate RDBMS, not a proprietary database

3. Support for dynamic, not precomputed, query environment

4. Requirement to store both detail data and summary data in the same physical database

5. Support for standards, including SQL, ODBC, and OLE DB

6. Portability among many platforms

7. Compatibility with a large number of third-party vendors

Strengths of Multidimensional Target Database

1. High performance, sophisticated multidimensional calculations

2. Optimized for OLAP; not constrained by limitations for SQL

3. Appropriate for many DSS applications, such as:

   ▼ Multiuser read/write applications; for example, budgeting analysis

   ▼ Support of complex, cross-dimensional calculations

   ▼ Drill-down for iterative queries, trend analysis, what-if analysis

4. Features supporting high performance for DSS calculations:

   ▼ High-speed multidimensional analysis and calculations; for example, aggregations, matrix calculations, cross-dimensional calculations, OLAP-aware functions, and procedural calculations

   ▼ Proprietary, extended features, such as row-level or matrix calculations, which are not supported by SQL

Limitations of Multidimensional Target Database

1. Requires yet another proprietary database solution, which increases the cost of training and support.

2. Maximum size of about three gigabytes of raw data, which may expand by a factor of 10 to 100 for calculated data. The size limitation is due to the amount of time required to precompute aggregates during data update window.

3. Most MDBs cannot load data incrementally.

4. Some MDBs require precomputation of all data — no on-the-fly calculations. This may result in a data explosion for sparse data.

5. Static, precomputed dimensions and calculations; lack of support for dynamic changes in dimensional structure

6. Performance degradation if the database size increases to more than 30 gigabytes

7. Lack of standard multidimensional model or access method

Hybrid OLAP

1. Combination of RDBMS and MDB, controlled by OLAP server:

   ▼ RDBMS is used for detailed data stored in large databases.

   ▼ MDB is used for fast, read/write OLAP analysis and calculations.

   ▼ OLAP server routes queries first to MDB and then to RDBMS.

   ▼ The result set from RDBMS may be processed on the fly in the server.

2. Representative hybrid and pseudo hybrid products:

▼ Microsoft SQL Server OLAP Services (Plato) launched January 1999

▼ Whitelight OLAP from Sybase (reseller)

▼ Oracle Express Server with ROLAP Option

▼ Holos from Seagate Software, Inc.

▼ OBM DB2 OLAP Server

Recommendations on Target Databases for Data Warehouses

Use RDBMS, MDB, and hybrid databases to meet the specialized requirements of groups of end users.

Use RDBMSs and ROLAP tools to provide multidimensional views of large relational target databases. Use RDBMS features, such as parallelization and bitmapped indexes to provide acceptable performance.

Use MDB for high-performance analysis of moderate-size databases. MDBs are often used for data marts.

Use the hybrid approach for applications that require access to detail data and fast OLAP computation.

DATA ACCESS AND ANALYSIS TOOLS

You have a wide choice in the selection of data access tools:

- Hybrid products that use the best features of both relational and multidimensional technologies

- High performance and scalability to large numbers of users

- Ease of use – intuitive, consistent interface to all functions of the tool

- Short learning curve – less than one hour for a desktop OLAP tool

- Specification of queries and reports without help from IS

- Semantic layer that maps business terms to database scheme

- Support for query, reporting, and OLAP analysis in same tool

- Wide range of price points (high-cost products for developers, moderate cost for LAN users, low cost for users on the Web)

- Aggregate navigation – ability to access precomputed aggregate

- Tight integration with central metadata repository.

- Conformance of data dimensions and attributes across data marts

- Support of *ad hoc* queries to any data in the data warehouse

- Rapid slicing across dimensions and attributes

- Support "drill anywhere" and "drill through" to detailed data

- Support for complex OLAP calculations and calculations on the fly

- Dynamic OLAP analysis capability over the Web

- Support for mobile computing, including off-line OLAP analysis

- Support of standards (Windows GUI, SQL, ODBC, OLE, OLE DB, OLE DB for OLAP, OIM, ISAPI, NSAPI, and CGI)

DESKTOP/RELATIONAL/MULTIDIMENSIONAL/ HYBRID OLAP TOOLS

1. Ease of use, intuitive interface, rapid response to ad hoc queries

2. Competitive price curve, including low-cost per user on the Web

3. Scalable architecture supporting large numbers of end users

4. Efficient distribution of functions across a three-tier architecture

5. OLAP calculation engine based on MDX and persistent, multidimensional cache on mid-tier server; local data cache on client

6. Support for both broadcasting and exception reporting

7. Multiuser, concurrent write capability

8. Ability to load data from a wide variety of sources

9. Close integration with RDBMSs to support hybrid architecture

10. Handle sparse and dense data without data explosion

Recommendations of Data Access and Analysis Tools

1. Use mix of tools to support multiple types of users:

   ▼ Managed query/reporting tools (optional)

   ▼ Desktop OLAP tools (support 70-80 percent of users)

▼ High-end ROLAP tools (support 10 percent of users)

▼ Multidimensional OLAP (MOLAP) tools (5-10 percent of users)

▼ Data mining tools (5 percent of users)

▼ Data visualization tools (5 percent of users)

2. Start with simple desktop OLAP tools. Gradually add more complex ROPLAP, MOLAP, and data mining tools.

LESSONS LEARNED

1. Make sure that the data warehouse is business driven, not technology driven.

2. Do not develop "virtual" data warehouses.

3. Do not develop "stovepipe" data marts that are not integrated across business areas.

4. Do not populate a data warehouse with "dirty" source data.

5. Do not implement enterprise data warehouse as a single, large, top-down development effort.

6. Anticipate scalability and performance issues.

Index

A

acquisition 186
aggregated fact table 46
algorithm 166, 223
algorithmic differential 85
alias 90
analysis session 62, 68
analyst, DSS 8
application layer 234
application links 179
application-level metadata 84
application-oriented data 4
array 204

B

backup 11
Bill Inmon 2, 242
binary cluster 155
bottom-up strategy 251

budgeting 240
budgeting process 240
business collaboration 244
business footprint 128
business intelligence 121, 153, 169, 171, 182, 238, 241, 242, 243, 244
business intelligence applications 238
business intelligence systems 238
business rules 144

C

central metadata repository 251
chain of abstraction 79
channel optimization 140
classification 165
clickstream data 104, 108, 129

EXPLORING LANS FOR THE SMALL BUSINESS & HOME OFFICE

Author: LOUIS COLUMBUS
ISBN: 0790612291 ● **SAMS#:** 61229
Pages: 304 ● **Category:** Computer Technology
Case qty: TBD ● **Binding:** Paperback
Price: $39.95 US/$63.95CAN

About the book: Part of Sams Connectivity Series, *Exploring LANs for the Small Business and Home Office* covers everything from the fundamentals of small business and home-based LANs to choosing appropriate cabling systems. Columbus puts his knowledge of computer systems to work, helping entrepreneurs set up a system to fit their needs.

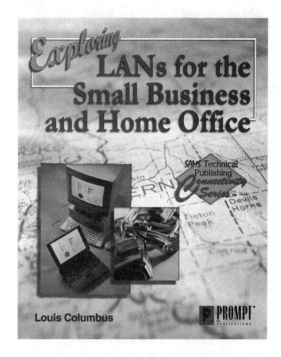

PROMPT® Pointers: Includes small business and home-office Local Area Network examples. Covers cabling issues. Discusses options for specific situations. Includes TCP/IP (Transmission Control Protocol/Internet Protocol) coverage. Coverage of protocols and layering.

Related Titles: *Administrator's Guide to E-Commerce*, by Louis Columbus, ISBN 0790611872. *Administrator's Guide to Servers*, by Louis Columbus, ISBN

Author Information: Louis Columbus has over 15 years of experience working for computer-related companies. He has published 10 books related to computers and has published numerous articles in magazines such as *Desktop Engineering, Selling NT Solutions*, and *Windows NT Solutions*. Louis resides in Orange, Calif.

To order today or locate your nearest PROMPT® Publications distributor at 1-800-428-7267 or www.samswebsite.com

Prices subject to change.

EXPLORING MICROSOFT OFFICE XP

Authors: JOHN BREEDEN & MICHAEL CHEEK
ISBN: 079061233X ● **SAMS#:** 61233
Pages: 336 ● **Category:** Computer Technology
Case qty: TBD ● **Binding:** Paperback
Price: $29.95 US/$47.95CAN
About the book: Breeden and Cheek provide an insight into the newest product from Microsoft — Office XP. Office XP is the replacement for Microsoft Office, designed to take users into the 21st century. Breeden and Cheek provide tips and tricks for the experienced office user, to help them find maximum value in this new software.

ELECTRONICS FOR THE ELECTRICIAN

Author: NEWTON C. BRAGA
ISBN: 0790612186 ● **SAMS#:** 61218
Pages: 320 ● **Category:** Electrical Technology
Case qty: 32 ● **Binding:** Paperback
Price: $34.95 US/$55.95CAN
About the book: Author Newton Braga takes an innovative approach to helping the electrician advance his or her career. Electronics have become more and more common in the world of the electrician, and this book will help the electrician become more comfortable and proficient at tackling the new tasks required.

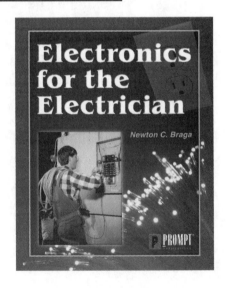

**To order today or locate your nearest PROMPT® Publications
distributor at 1-800-428-7267 or www.samswebsite.com**

Prices subject to change.

SEMICONDUCTOR CROSS REFERENCE BOOK, 5/E

Author: SAMS TECHNICAL PUBLISHING
ISBN: 0790611392 ● **SAMS#:** 61139
Pages: 876 ● **Category:** Professional Reference
Case qty: 14 ● **Binding:** Paperback
Price: $39.95 US/$63.95CAN
About the book: The perfect companion for anyone involved in electronics! Sams has compiled years of information to help you make the most of your stock of semiconductors. Both paper and CD-ROM versions of this tool contain an additional 128,000 parts listings over the previous editions.

ON CD-ROM, 2E
ISBN: 0790612313 ● **SAMS#:** 61231 ● **Price:** $39.95 US/$63.95CAN

COMPUTER NETWORKS FOR THE SMALL BUSINESS & HOME OFFICE

Author: JOHN A. ROSS
ISBN: 0790612216 ● **SAMS#:** 61221
Pages: 304 ● **Category:** Computer Technology
Binding: Paperback ● **Price:** $39.95 US/$63.95CAN
About the book: Small businesses, home offices, and satellite offices with unique networks of 2 or more PCs can be a challenge for any technician. This book provides information so that technicians can install, maintain and service computer networks typically used in a small business setting. Schematics, graphics and photographs will aid the "everyday" text in outlining how computer network technology operates, the differences between various network solutions, hardware applications, and more.

To order today or locate your nearest PROMPT® Publications distributor at 1-800-428-7267 or www.samswebsite.com

Prices subject to change.

APPLIED ROBOTICS II

Author: EDWIN WISE
ISBN: 0790612224 • **SAMS#:** 61222
Pages: 304 • **Category:** Projects
Case qty: TBD • **Binding:** Paperback
Price: $29.95 US/$47.95CAN

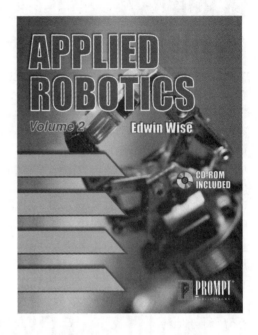

About the book: Edwin Wise builds upon his best-seller, *Applied Robotics* with this book targeted at more advanced hobbyists with development of a larger, more robust, and very practical mobile robot platform. Building on the foundation set in his first text, *Applied Robotics II* has projects to create a larger robot platform suitable for use in the home or outdoors, advanced sensor projects and a great exploration of A1 and control software.

Prompt Pointers: Picks up where *Applied Robotics* left off. Offers an advanced set of projects related to this very hot subject area.

Related Titles: *Applied Robotics*, ISBN 0790611848. *Animatronics*, ISBN 079061294.

Author Information: Edwin Wise is a professional software engineer with twenty years of experience. He currently works in the field of Computer Aided Manufacturing (CAM). His experience includes work on both computer games and educational software. Building robots has been a dream and passion for Edwin for years now. His current project is "Boris," a giant killer robot that can be viewed at http://www.simreal.com/Boris.

GUIDE TO CABLING AND COMMUNICATION WIRING

Author: LOUIS COLUMBUS
ISBN: 0790612038 ● **SAMS#:** 61203
Pages: 320 ● **Category:** Communications
Case qty: TBD ● **Binding:** Paperback
Price: $39.95 US/$63.95CAN

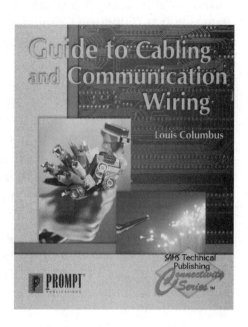

About the book: Part of Sams Connectivity Series, *Guide to Cabing and Communication Wiring* takes the reader through all the necessary information for wiring networks and offices for optimal performance. Columbus goes into LANs (Local Area Networks), WANs (Wide Area Networks), wiring standards and planning and design issues to make this an irreplaceble text.

PROMPT® Pointers:
Features planning and design discussion for network and telecommunications applications. Explores data transmission media. Covers Packet Framed-based data transmission.

Related Titles: *Administrator's Guide to E-Commerce*, by Louis Columbus, ISBN 0790611872. *Exploring LANs for the Small Business and Home Office*, by Louis Columbus, ISBN 0790612291. *Computer Networking for the Small Business and Home Office*, by John Ross, ISBN 0790612216.

Author Information: Louis Columbus has over 15 years of experience working for computer-related companies. He has published 10 books related to computers and has published numerous articles in magazines such as *Desktop Engineering, Selling NT Solutions*, and *Windows NT Solutions*. Louis resides in Orange, Calif.

PC HARDWARE: UPGRADING AND REPAIR

Author: MICHAEL GRAVES
ISBN: 079061250X • **SAMS#:** 61250
Pages: 800 • **Category:** Computer Technology
Case qty: TBD • **Binding:** Paperback
Price: $39.95 US/$63.95CAN

About the book: As the technology surrounding our desktop PCs continues to evolve at a rapid pace, the opportunity to understand, repair and upgrade your PC is attractive. In an era where the PC you bought last year is now "out of date", your opportunity to bring your PC up-to-date rests in this informative text. Renouned author Michael Graves addresses this subject in a one-on-one manner, explaining each category of computer hardware in a complete, concise manner.

Prompt Pointers: Designed to bring a beginner up to a professional level of hardware expertise. Includes new SCSI III implementations, new video standards, and previews of upcoming technologies.

Related Titles: *Exploring Office XP*, ISBN 079061233X, *Designing Serial SANS*, ISBN 0790612461, *Administrator's Guide to Datawarehousing*, ISBN 0790612496.

Author Information: Michael Graves is a Senior Hardware Technician and Network Engineer for Panurgy of Vermont. Graves has taught computer hardware courses on the college level at Champlain College in Burlington, Vermont and The Essex Technical Center in Essex Junction, Vermont. While this is his first full-length book under his own name, his contributions have been included in other works and his technical writing has been the source of several of the more readable user's guides and manuals for different products.

AUTOMOTIVE AUDIO SYSTEMS

Author: HOMER L. DAVIDSON
ISBN: 0790612356 ● **SAMS#:** 61235
Pages: 320 ● **Category:** Automotive
Case qty: TBD ● **Binding:** Paperback
Price: $39.95 US/$63.95CAN

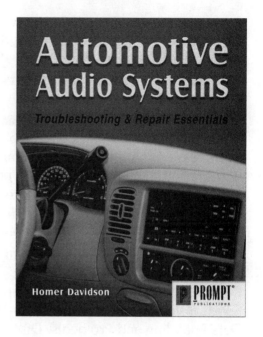

About the book: High-powered car audio systems are very popular with today's under-30 generation. These top-end systems are merely a component within the vehicle's audio system, much as your stereo receiver is a component of your home audio and theater system. Little has been written about the troubleshooting and repair of these very expensive automotive audio systems. Homer Davidson takes his decades of experience as an electronics repair technician and demonstrates the ins-and-outs of these very high-tech components.

Prompt Pointers: Coverage includes repair of CD, Cassette, Antique car radios and more. All of today's high-end components are covered. Designed for anyone with electronics repair experience.

Related Titles: *Automotive Electrical Systems*, ISBN 0790611422. *Digital Audio Dictionary*, ISBN 0790612011. *Modern Electronics Soldering Techniques*, ISBN 0790611996.

Author Information: Homer L. Davidson worked as an electrician and small appliance technician before entering World War II teaching Radar while in the service. After the war, he owned and operated his own radio and TV repair shop for 38 years. He is the author of more than 43 books for TAB/McGraw-Hill and Prompt Publications. His first magazine article was printed in *Radio Craft* in 1940. Since that time, Davidson has had more than 1000 articles printed in 48 different magazines. He currently is TV Servicing Consultant *for Electronic Servicing & Technology* and Contributing Editor for *Electronic Handbook*.

To order today or locate your nearest PROMPT® Publications distributor at 1-800-428-7267 or www.samswebsite.com

Prices subject to change.

SERIAL SANS

Author: WILLIAM DAVID SCHWADERER
ISBN: 0790612461 ● **SAMS#:** 61246
Pages: 320 ● **Category:** Computer Technology
Case qty: TBD ● **Binding:** Paperback
Price: $39.95 US/$63.95CAN

About the book: The use of Serial SANS is an increasingly popular and efficient way to store data in a medium to large corporation setting. Serial SANS effectively stores your company's data away from the traditional server, allowing your valuable server resources to be used for running applications.

Prompt Pointers: Covers Device Specialization Considerations. Explains Media Signals, Data Encoding and Protocols. Discusses SAN hardware building blocks.

Related Titles: *Administrator's Guide to Datawarehousing*, ISBN 0790612496, *How the PC Hardware Works*, ISBN 079061250X.

Author Information: W. David Schwaderer has extensive complex computer system experience and was involved in the creationof two Silicon Valley start-up companies. Schwaderer has a diverse background in connectivity products, personal computer software, and voice DSP based systems. Schwaderer currently resides in Saratoga, Calif.

EXTRANET/INTRANET

Author: CONRAD PERSSON
ISBN: 0790612410 ● **SAMS#:** 61241
Pages: 304 ● **Category:** Computer Technology
Case qty: TBD ● **Binding:** Paperback
Price: $34.95 US/$55.95CAN

About the book: We are all familiar with the Internet, but few of us have occasion to utilize an Intranet or Extranet application. Both have vast applications related to inner-company communication, customer service, and vendor relations. Both are built similarl to Internet sites, and have many of the same features, issues, and problems. Intranet and Extranet applications are generally under-utilized, even though they provide the opportunity for both communication and financial benefits.

Prompt Pointers: Designed for the Systems Administrator or advanced webmaster. Outlines Intranet/Extranet issues, problems, and opportunities. Discusses hardware and software needs.

Related Titles: *Administrators Guide to E-Commerce,* ISBN 0790611872. *Computer Networking for the Small Business and Home Office,* ISBN 0790612216. *Exploring Microsoft Office XP,* ISBN 079061233X.

Author Information: Conrad Persson is the editor of *ES&T Magazine,* the premier publication for the electronics servicing industry. Conrad has decades of experience related to electronics and computer applications and resides in Shawnee Mission, Kan.

BROADBAND

Author: MICHAEL BUSBY
ISBN: 0790612488 ● **SAMS#:** 61248
Pages: 352 ● **Category:** Communications
Case qty: TBD ● **Binding:** Paperback
Price: $39.95 US/$63.95CAN

About the book: As the telecommunications industry goes through deregulation, the lines between telecom and other communication applications have become very blurred. Business mergers and advancing technology have created a need for more and more broadband applications. Author Michael Busby addresses these issues in relation to the telecommunications sector as well as topics pertaining to cable, satellite, RF, microwave and other communication methods. A must read for anyone working with telecommunication technologies.

Prompt Pointers: Includes discussions of LAN, CAD, imaging, wire, cable and more. Addresses networking fundamentals, protocols, and multimedia applications.

Related Titles: *Telecommunication Technologies*, ISBN 0790612259, *Exploring LANS for the Small Business & Home Office*, ISBN 0790612291, *Guide to Cabling & Communication Wiring*, ISBN 0790612038.

Author Information: Michael Busby is president and CEO of Mikal Enterprises, a global telecommunications design and consulting company.Busby has over 30 years telecommunications experience as field service engineer, systems engineer, R&D engineer, engineering manager, product manager, and VP engineering.

To order today or locate your nearest PROMPT® Publications distributor at 1-800-428-7267 or www.samswebsite.com

Prices subject to change.

HOME THEATER SYSTEMS

Author: BOB GOODMAN
ISBN: 0790612372 ● **SAMS#:** 61237
Pages: 304 ● **Category:** Video Technology
Case qty: TBD ● **Binding:** Paperback
Price: $39.95 US/$63.95CAN

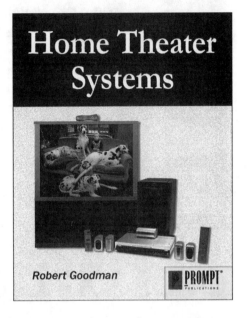

About the book: In days past, you had a TV, radio, and maybe a turntable in your "living room." Today, the evolution of electronics has brought us the Home Theater System, combining projection TVs, high-powered audio receivers, multiple CD players, DVD systems, surround-sound and more. This plethora of components is rarely purchased from a single manufacturer, making installation and maintenance a complicated task at best. Bob Goodman applies his electronics experience to this topic and provides a guidebook to home theater systems, including information on systems, components, troubleshooting, and maintenance.

Prompt Pointers: Home theater systems are the future of home audio/video systems. A buyer's guide is included. Great detail is provided regarding component choices.

Related Titles: *Digital Audio Dictionary*, ISBN 0790612011. *DVD Player Fundamentals*, ISBN 0790611945. *Guide to Satellite TV Technology*, ISBN 0790611767.

Author Information: Bob Goodman, CET, has devoted much of his career to developing and writing about more effective, efficient ways to troubleshoot electronics equipment. An author of more than 62 technical books and 150 technical articles, Goodman spends his time as a consultant and lecturer in Western Arkansas.

To order today or locate your nearest PROMPT® Publications distributor at 1-800-428-7267 or www.samswebsite.com

Prices subject to change.

BASIC SOLID STATE ELECTRONICS

Author: VAN VALKENBURG
ISBN: 0790610426 ● **SAMS#:** 61042
Pages: 944 ● **Category:** Electronics Technology
Case qty: 12 ● **Binding:** Paperback
Price: $29.95 US/$47.95CAN
About the book: Considered to be one of the best books on solid-state electronics on the market, this revised edition provides the reader with a progressive understanding of the elements that form various electronic systems. Electronic fundamentals covered in the illustrated, easy-to-understand text include semiconductors, power supplies, audio and video amplifiers, transmitters, receivers, and more.

CMOS SOURCEBOOK

Author: NEWTON BRAGA
ISBN: 0790612348 ● **SAMS#:** 61234
Pages: 304 ● **Category:** Electronics Technology
Case qty: TBD ● **Binding:** Paperback
Price: $39.95 US/$63.95CAN
About the book: CMOS (Complementary Metal Oxide Semiconductors) are an essential part of almost every electronics component and are not typically understood. Braga takes the concepts from the legendary CMOS Cookbook from Don Lancaster (originally published by Sams/ Macmillan) and brings them into the 21st Century with this new and different look at CMOS IC technology.

To order today or locate your nearest PROMPT® Publications distributor at 1-800-428-7267 or www.samswebsite.com

Prices subject to change.

ADMINISTRATOR'S GUIDE TO E-COMMERCE

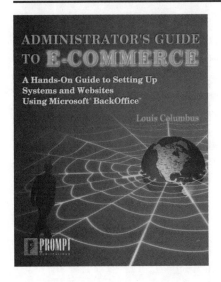

Author: LOUIS COLUMBUS
ISBN: 0790611872 ● **SAMS#:** 61187
Pages: 416 ● **Category:** Business Technology
Case qty: 28 ● **Binding:** Paperback
Price: $34.95 US/$55.95CAN
About the book: Unlike previous electronic commerce books which stress theory, the Administrator's Guide to E-Commerce is a hands-on guide to creating and managing Web sites using the Microsoft BackOffice product suite. This book will explore the role of networking technologies to industry growth, issues of privacy and security, and most importantly, guidance in taking an existing Web server and creating an electronic storefront.

ANIMATRONICS: GUIDE TO HOLIDAY DISPLAYS

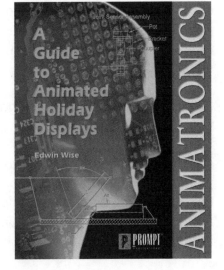

Author: EDWIN WISE
ISBN: 0790612194 ● **SAMS#:** 61219
Pages: 304 ● **Category:** Projects
Case qty: 32 ● **Binding:** Paperback
Price: $29.95 US/$47.95CAN
About the book: Author Edwin Wise takes the reader inside his world of robotics in an innovative guide to designing, developing, and building animated displays centered around the holidays of Halloween and Christmas.

To order today or locate your nearest PROMPT® Publications distributor at 1-800-428-7267 or www.samswebsite.com

Prices subject to change.

APPLIED ELECTRONIC SURVEILLANCE DEVICES & CIRCUITS

Author: CARL BERGQUIST
ISBN: 0790612453 ● **SAMS#:** 61245
Pages: 304 ● **Category:** Video Technology
Case qty: TBD ● **Binding:** Paperback
Price: $32.95 US/$52.50CAN

About the book: One of the hottest topics of 2001, Surveillance, is covered in dept in this text from Carl Bergquist. The accessability of electronic components, Internet access and affordable parts coupled with the increased fears and concerns of the general public has created a boom in the surveillance industry. Bergquist covers the building of surveillance systems including video surveillance, wireless systems, computer network systems, and audio systems. Also discussed are issues related to this sensitive topic, uses, legal considerations, counter-surveillance and much more.

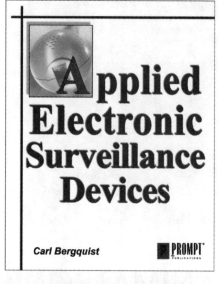

Prompt Pointers: One of the hottest electronic topics of 2001! Includes 6 do-it-yourself projects. Designed for anyone with electronics knowledge.

Author Information: Carl Bergquist followed a successful career as a photojournalist for AP, UPI, *The New York Times*, *Newsweek*, and other publications by turning his efforts toward a lifelong hobby of electronics. Besides articles in *Popular Electronics* and *Electronics Now*, Carl has authored numerous books for Prompt Publications.

To order today or locate your nearest PROMPT® Publications distributor at 1-800-428-7267 or www.samswebsite.com

Prices subject to change.

BASIC ELECTRICITY

Author: VAN VALKENBURG
ISBN: 0790610418 ● **SAMS#:** 61041
Pages: 736 ● **Category:** Electrical Technology
Case qty: 16 ● **Binding:** Paperback
Price: $29.95 US/$47.95CAN
About the book: Considered to be one of the best electricity books on the market, the authors have provided a clear understanding of how electricity is produced, measured, controlled and used. A minimum of mathematics is used for direct explanations of primary cells, magnetism, Ohm's Law, capacitance, transformers, DC generators, and AC motors. Other essential topics covered include conductance, current flow, electromagnetism and meters.

Best Seller!

BASIC ELECTRICITY AND DC CIRCUITS

Author: CHARLES DALE
ISBN: 0790610728 ● **SAMS#:** 61072
Pages: 928 ● **Category:** Electronics Technology
Case qty: 10 ● **Binding:** Paperback
Price: $39.95 US/$63.95CAN
About the book: No matter what their background, readers can learn the basic concepts that have enabled mankind to harness and control electricity. Chapters are arranged to allow readers to progress at their own pace, with concepts and terms being introduced as needed for comprehension.

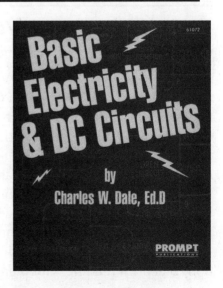

To order today or locate your nearest PROMPT® Publications distributor at 1-800-428-7267 or www.samswebsite.com

Prices subject to change.

HOME AUTOMATION BASICS - PRACTICAL APPLICATIONS USING VISUAL BASIC 6

Author: THOMAS LEONIK
ISBN: 0790612143 ● **SAMS#:** 61214
Pages: 386 ● **Category:** Projects ● **Binding:** Paper
Price: $34.95 US/$55.95CAN
About the book: This book explores the world of Visual Basic 6 programming with respect to real-world interfacing and control on a beginner to intermediate level, with a home automation system. Includes HVAC systems, water pumps, temperature controls and more.

HOME AUTOMATION BASICS II - LITETOUCH SYSTEM

Author: JAMES VAN LAARHOVEN
ISBN: 0790612267 ● **SAMS#:** 61226
Pages: 336 ● **Category:** Projects
Case qty: 32 ● **Binding:** Paperback
Price: $34.95 US/$55.95CAN
About the book: James Van Laarhoven explores the very comprehensive home automation system from LiteTouch Systems. This book will aid in the installation, maintenance and programming of the LiteTouch 2000. Includes lighting, audio, installation, blueprint reading, video and more.

To order today or locate your nearest PROMPT® Publications distributor at 1-800-428-7267 or www.samswebsite.com

Prices subject to change.

GUIDE TO DIGITAL CAMERAS

Author: MICHAEL MURIE
ISBN: 0790611759 ● **SAMS#:** 61175
Pages: 536 ● **Category:** Video Technology
Case qty: 18 ● **Binding:** Paperback
Price: $39.95 US/$63.95CAN
About the book: The Complete Guide to Digital Cameras will appeal to anyone who has recently purchased or is considering an investment in a digital camera. Together the book and CD-ROM will answer questions you have about digital cameras, enable you to make intelligent buying decisions, and help you use your camera to its full potential. No camera purchase is complete without this informative guide.

GUIDE TO WEBCAMS

Author: JOHN BREEDEN & JASON BYRNE
ISBN: 0790612208 ● **SAMS#:** 61220
Pages: 320 ● **Category:** Video Technology
Binding: Paperback
Price: $29.95 US/$47.95CAN
About the book: Digital video cameras have become an increasingly popular method of communicating with others across the Internet. From video e-mail clips to websites that broadcast people's lives for all to see, webcams have become a way for people to battle against the perceived threat of depersonalization caused by computers and to monitor areas from anywhere in the world.

**To order today or locate your nearest PROMPT® Publications
distributor at 1-800-428-7267 or www.samswebsite.com**

Prices subject to change.

DSP FILTER COOKBOOK

DSP Filters

John Lane
Jayant Datta
Brent Karley
Jay Norwood

Electronics Cookbook Series

Author: JOHN LANE, ET AL
ISBN: 0790612046 ● **SAMS#:** 61204
Pages: 344 ● **Category:** Electronics Technology
Case qty: 26 ● **Binding:** Paperback
Price: $39.95 US/$63.95CAN
About the book: Digital filters and real-time processing of digital signals have traditionally been beyond the reach of most, due partially to hardware cost as well as complexity of design. In recent years, low-cost digital signal processor (DSP) development boards have put this within reach. This book will break down this design complexity barrier by means of simplified tutorials, step-by-step instructions, along with a collection of audio projects.

TELECOMMUNICATION TECHNOLOGIES

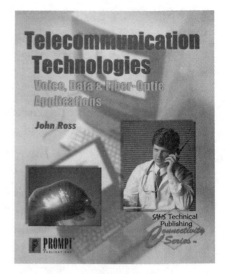

Author: JOHN A. ROSS
ISBN: 0790612259 ● **SAMS#:** 61225
Pages: 368 ● **Category:** Communications
Binding: Paperback
Price: $39.95 US/$63.95CAN
About the book: This book contains the information needed to develop a complete understanding of the technologies used within telephony, data and telecommunications networks.
Projects and topics include: Equipment comparisons, business office applications and understanding the technology.

To order today or locate your nearest PROMPT® Publications distributor at 1-800-428-7267 or www.samswebsite.com

Prices subject to change.